HAMBURG
ELSFLETH
DEUTSCHLAND
AÇORES
ISLAS
CANARIAS
SANTA CRUZ DE TENERIFE
PUERTO DE MOGÁN
ER
SÉNÉGAL
CÔTE
D'IVOIRE
KEY
SHIP'S ROUTE
EARLIER ROUTE
OCEAN CURRENT
PREVAILING WIND
D1600083

'This book is both important and beautiful: important, in that it describes one of the best ways we can move into a post-fossil fuels civilisation, which is to say by sail; and beautiful, because it shows on every page how this bursting out of the cocoon of heavy oil that we have been living in will return us to a life in the real world, with the wind felt in the hands and on one's face, and every day an adventure. What a joy to read these pages and learn their news.'

Kim Stanley Robinson, author of
The Ministry for the Future

'The story is original – a sailing ship undertaking a quixotic mission to deliver a tiny amount of cargo (in order to demonstrate that it can be done) when the pandemic descends, trapping the narrator – makes for a diverting narrative.'

Horatio Clare, author of
Down to the Sea in Ships

'This is a book that should change the world. Beautifully written and brimming with bold yet careful analysis, Christiaan De Beukelaer has given the world a tremendous gift.'

Deborah Cowen, author of
The Deadly Life of Logistics

'De Beukelaer tackles a subject that's growing in importance, giving us his perspective from the deck of a sailing cargo boat. No mean achievement.'

Tom Cunliffe, sailor, presenter and author of
The Complete Yachtmaster

‘*Trade winds* is an absorbing account of a voyage that starts off as an effort to prove the continuing viability of sail, but becomes far more challenging than expected when the COVID-19 pandemic shuts off all access to the shore. It is also a thoughtful analysis of practical ways for shrinking the carbon footprint of one of the world’s most polluting industries – shipping.’

Amitav Ghosh, author of *The Nutmeg’s Curse*

‘A truly fascinating account – of a voyage, but also of an idea that is counter-intuitive in a world based on speed, but revelatory for a planet that is going to have to start taking real care of itself. There’s a bit of romance here, and a lot of reality.’

Bill McKibben, author of *The End of Nature*

‘Simultaneously engaging and scholarly, *Trade winds* combines sailing memoir and environmental analysis to provide important insights about the environmental effects of global shipping and about what plausible alternatives might be. De Beukelaer brings together personal experience with an impressive command of scholarly literatures across history, economics, philosophy, climate science, among many others, and brings them all together in an interdisciplinary *tour de force* that is realistic and yet ultimately hopeful.’

Elizabeth R. DeSombre, Camilla Chandler Frost Professor of Environmental Studies, Wellesley College

‘Scientists, scholars, and sailors have long turned to the ocean to conjure futures that lie just beyond reach. In his dive into the quixotic movement to revive shipping by sail, Christiaan De Beukelaer leaves us wondering which is more of a folly: imagining that sail could once again be a viable technology for transporting goods around the world, or imagining that we can continue practicing business as usual.’

Philip Steinberg, UArctic Chair in Political Geography, Durham University

'Finally a book that delves deeply into the urgent need to decarbonise the shipping industry by connecting political economy and environmental challenges. Christiaan De Beukelaer poignantly describes day-to-day life aboard a twenty-first-century sail cargo vessel in its social and historical context. In doing so, this book raises essential questions about the future of shipping while offering suggestions on how to resolve them. This story of adventure on the high seas sketches a liminal space that will inspire realists and dreamers alike.'

Dr Lucy Gilliam, Senior Shipping Policy Officer, Seas at Risk

'The decarbonisation of shipping does not need to be a sacrifice. Instead, as Christiaan De Beukelaer eloquently shows, sailing to a sustainable future for shipping can be exciting and full of enriching experiences for both the author and us, the readers. Admittedly, I personally do not share all the views of Christiaan about open registries and the workings of the maritime industry, but I appreciate his sincere and enthusiastic voyage. I recommend reading this thought-provoking book.'

Jan Hoffman, Chief of UNCTAD's Trade Logistics Branch

'*Trade winds* is a riveting book that talks about seafarers' workers' rights, sustainable trade that's fair and equitable, the human struggles of the lockdown and the COVID pandemic, the climate crisis, and the emissions of the shipping industry. It says all this in a beautiful story that emphasizes the need for storytelling and radically imagining the better world that we're fighting for. It puts practical solutions to take our first steps towards climate justice while simultaneously challenging us to think about how and where we're going next.'

Mitzi Jonelle Tan, Youth Advocates for Climate Action Philippines & Fridays for Future

'This book brings the environmental and social challenges facing the shipping industry to the fore. After five months at sea, Christiaan reveals how the radical changes necessary to decarbonise shipping will rely on the skill, tenacity and sacrifices of seafarers. This book rightly asks the big question, how do we build an economically, socially and environmentally sustainable shipping industry? An important contribution to the challenges facing shipping that exposes the work needed to ensure the fundamental rights of seafarers are protected.'

Stephen Cotton, General Secretary, International Transport Workers' Federation (ITF)

'As the famous quote goes: "If you want to build a ship, don't drum up the men to gather wood, divide the work, and give orders. Instead, teach them to yearn for the vast and endless sea." This is a book that does just that! Christiaan's intense interest and vast knowledge on this subject matter jumps out from every page, whether it is directly recounting his extended sail cargo adventure in lively detail or using those experiences to bring alive the fabric and challenges of maritime transport and trade past and present. A story of sustainability, shipping and a glimpse of a maritime future full of possibilities that will sustain your interest throughout.'

Gavin Allwright, Secretary General of the International Windship Association (IWSA)

Trade winds

Manchester University Press

Trade winds

A voyage to a sustainable future for shipping

Christiaan De Beukelaer

Manchester University Press

Published by Manchester University Press
Oxford Road, Manchester M13 9PL
www.manchesteruniversitypress.co.uk

British Library Cataloguing-in-Publication Data
A catalogue record for this book is available from the British Library

ISBN 978 1 5261 6309 7 hardback

First published 2023

Typeset
by Cheshire Typesetting Ltd, Cuddington, Cheshire
Printed in Great Britain
by Bell & Bain Ltd, Glasgow

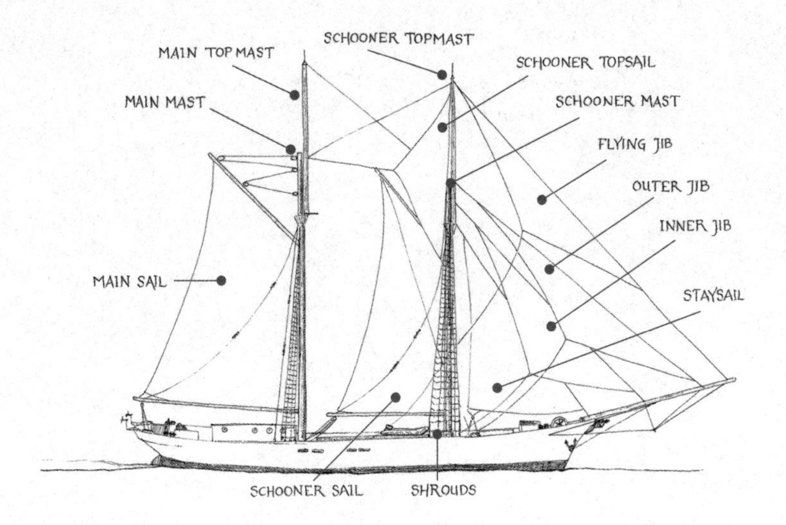
SCHOONER TOPMAST
MAIN TOP MAST
SCHOONER TOPSAIL
MAIN MAST
SCHOONER MAST
FLYING JIB
OUTER JIB
INNER JIB
MAIN SAIL
STAYSAIL
SCHOONER SAIL
SHROUDS

In loving memory of
Jean François 'Papoe' De Beukelaer

With illustrations by
Athena Corcoran-Tadd

And photographs by
Christiaan De Beukelaer

Contents

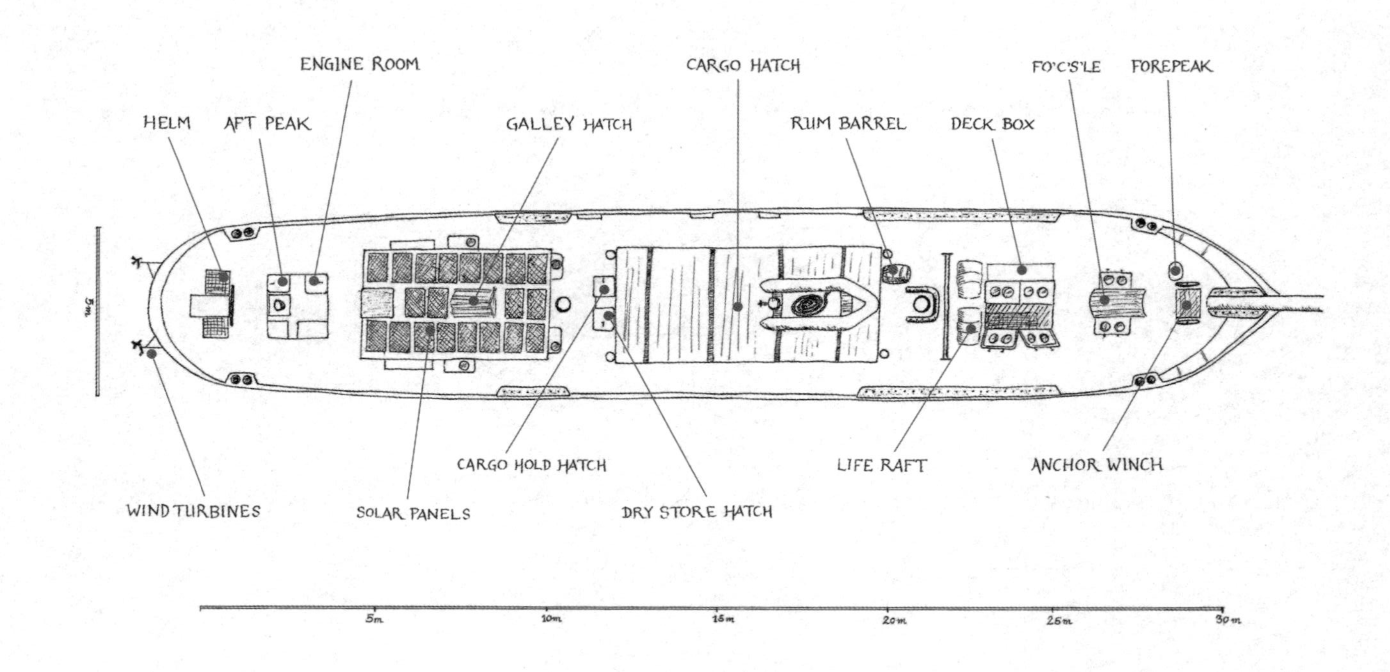
ENGINE ROOM
CARGO HATCH
FO'C'S'LE
FOREPEAK
HELM
AFT PEAK
GALLEY HATCH
RUM BARREL
DECK BOX
5m
CARGO HOLD HATCH
LIFE RAFT
ANCHOR WINCH
WIND TURBINES
SOLAR PANELS
DRY STORE HATCH
5m
10m
15m
20m
25m
30m

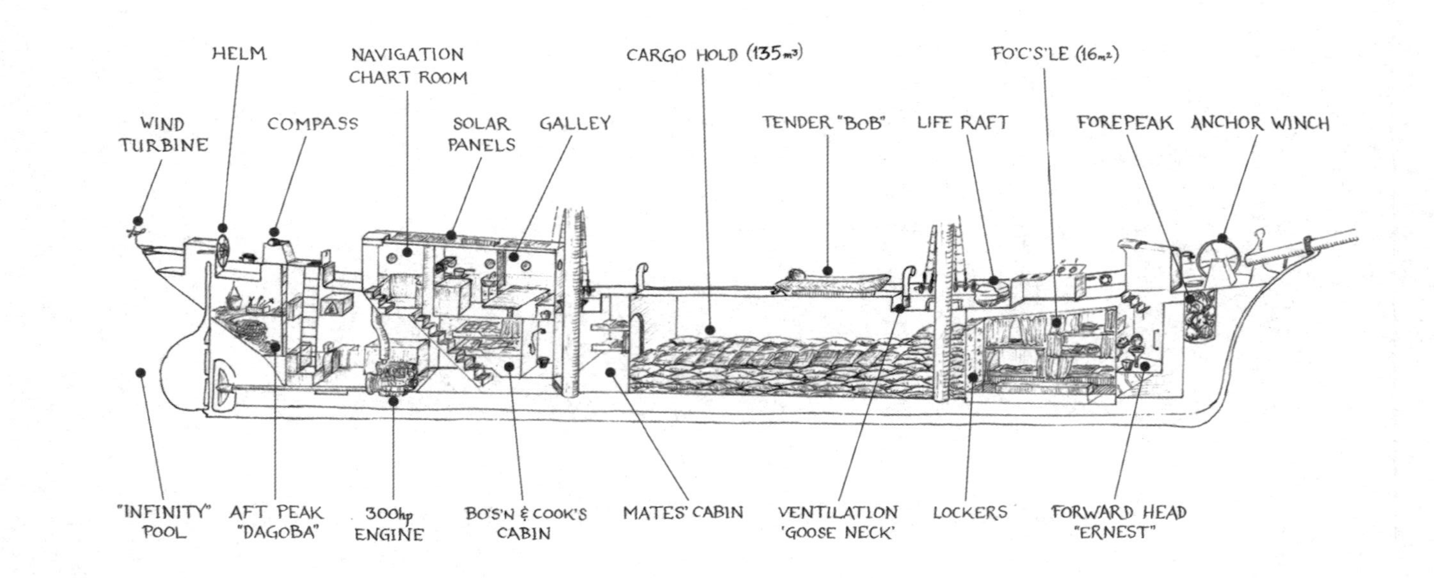

HELM
NAVIGATION CHART ROOM
CARGO HOLD ($135m^3$)
FO'C'S'LE ($16m^2$)
WIND TURBINE
COMPASS
SOLAR PANELS
GALLEY
TENDER "BOB"
LIFE RAFT
FOREPEAK
ANCHOR WINCH
"INFINITY" POOL
AFT PEAK "DAGOBA"
300hp ENGINE
BO'S'N & COOK'S CABIN
MATES' CABIN
VENTILATION 'GOOSE NECK'
LOCKERS
FORWARD HEAD "ERNEST"

Annus pandemicus

Where were you in 2020? What is your pandemic lockdown story?

These questions now mark our lives.

Did you get infected? Did you lose someone you know? How badly did the virus affect your livelihoods and life? What was the political reaction where you live? Did you get stranded overseas? Did you spend months indoors with your children? Or were you happily 'working from home'?

Melbourne, the city I call home, went into lockdown in late March 2020. By early June, the city was slowly opening up again. But I escaped it all. I was at sea. I spent countless days bobbing around the *Grand Bleu*, among the flying fish of the mid-Atlantic, the whales off Newfoundland, the frigate birds above the Gulf of Mexico, and dolphins that popped up and disappeared almost every day.

In many ways, living aboard a ship is like going into lockdown voluntarily. In the eighteenth century, the English writer Samuel Johnson compared being at sea to being locked up in jail, with the significant difference that you might drown at sea. Much like a prison, you cannot leave a ship at will, and though the chance of drowning is real, the view at sea is better.

The prison analogy gained pertinence as we were confined to the *Avontuur* and banned from making landfall. We watched the 1962 film *Mutiny on the Bounty* and it struck me that we were more like a band of mutineers than ordinary sailors. After the mutineers had sent their captain, Vice-Admiral William Bligh, packing in a launch, the protagonist and master's mate Fletcher Christian loudly disagreed with one of his fellow mutineers who argued that they had the choice between prison or mutiny. 'You're in prison now, Mills.' Fletcher stressed, 'With one difference. We are not locked in; we are locked out.'

Being at sea, we lived through the pandemic rather differently than those ashore. We had limited contact with the world; we were wholly outside the frenzy of the twenty-four-hour news cycle. We merely sailed, as any sailor would. Standing watch, trimming sail, helming, eating, cleaning, and sleeping.

I was aboard the *Avontuur* to understand her mission. The century-old two-masted schooner transports cargo across the Atlantic Ocean using the wind. Before the marquis de Jouffroy d'Abbans tested the *Pyroscaphe* on July 15, 1783, the world's first steam-powered ship, on the river Saône in France, every single vessel in the world used sails (or oars) for propulsion.

Today, owing to a range of factors, including the low price and abundance of oil, very few sailing vessels carry cargo anymore. Sailing has become a form of *sport* rather than *transport*. But the *Avontuur*, currently the largest sailing cargo vessel in the world, aims to change that. I joined her on the Canary Islands to pick up a cargo of rum, coffee, and cacao in the Caribbean and Mexico, destined for Germany. In doing so, I hoped to find out more about the revival of working sailing ships, which transport cargo using the wind as an emission-free means of propulsion.

This book is part travel story, part climate activism, part reflection on the past and future of the shipping industry. I wrote it because of the sheer urgency to tackle the carbon emissions of shipping. It is far from the only issue that plagues the oceans in the Anthropocene, which include rapidly diminishing marine biodiversity, enormous patches of drifting plastic, ocean acidification as the water absorbs carbon dioxide from the air, and melting sea ice that reduces the reflective capacity of the planet's surface, triggering further warming. These issues are urgent too, and many more could be added to the list. But in this book, I primarily focus on the challenge we face in changing the means of propulsion of maritime trade, away from polluting fossil fuels to zero-emission technologies and fuels. While most of the industry and the International Maritime Organization (IMO), its regulatory body, focus on those alternative fuels, my focus here is on the potential of sail – arguably the zero-emission technology par excellence – in maintaining global supply chains in our globalised economy, while ensuring that shipping doesn't wreck the planet.

History is awash with people who left their safe, shore-based lives to sail the seas. In 1834, Richard Henry Dana Jr dropped out of Harvard College to join the merchant marine – not as an officer, typical for someone with a privileged background, but as an ordinary seaman. He sailed on the *Pilgrim* and later the *Alert* from Boston to California, passing the notoriously rough Cape Horn. His book *Two Years Before the Mast* is a rare first-hand account of a seafarer before the mast – this term refers both to the location of ordinary seafarers' living quarters and the spatial demarcation of difference between (commonly rich) officers and (commonly working-class) ordinary seafarers. It gives us an insight into sailors' lives in the nineteenth century, the heyday of sailing cargo vessels. Fast clipper ships connected ports worldwide to supply keen buyers with precious goods like tea and opium – or in Dana's case, hides from cattle ranches in the Wild West.[1]

His story inspired Herman Melville to head to sea, which resulted in *Moby-Dick* (1851), an early icon of American literature. Dana inspired scores of others to set sail, despite his stern warning that life at sea wasn't the romantic idyll some imagined: 'A sailor's life is at best but a mixture of a little good with much evil, and a little pleasure with much pain. The beautiful is linked with the revolting, the sublime with the commonplace, and the solemn with the ludicrous.'[2]

A hundred years later, in 1938, Eric Newby quit his job at The Wurzel Agency, a London advertising firm, to follow in the footsteps of Dana and Melville. On the eve of the Second World War, he embarked on the *Moshulu* in Belfast to pick up grain in Port Victoria, South Australia, bound for Glasgow in Scotland. He called his book *The Last Grain Race*. While sailing cargo vessels had become rare by the time Newby set sail, he had no way of knowing that the Second World War would mean their end. 'War,' wrote the historian A. J. P. Taylor, has 'always been the mother of all invention.'[3] Unfortunately, war cares little for life or planet. The discovery of oil in Damman, Saudi Arabia, on March 3, 1938, did not help much either. By the time he published his book in 1956, it was clear that the last days of sail, as the Australian master mariner and writer Alan Villiers calls the period, were over.[4] Arguably, this is a recurrent theme in writing about sailing and the sea, as Herman Melville, Joseph Conrad, and Henry Dana Jr offer insights into disappearing maritime worlds rather than chronicling them in their heyday.[5]

In the early twenty-first century, sailing cargo vessels are making a surprising come-back as the world witnesses the devastating impact of climate change driven by the burning of fossil fuels. The first time I heard about their revival was when the *Tres Hombres* set sail in 2007. At the time, I was living in a squat in Amsterdam where my local pub was Joe's Garage, the legendary squatter's den in the city's working-class east – which was rapidly gentrifying at the time.[6] Over a beer or two, I caught wind that a ship was to start sailing goods across the Atlantic Ocean. This brigantine, launched in 1943, had a cargo capacity of 40 tonnes and was named after Jorne Langelaan, Andreas Lackner, and Arjen van der Veen, the three 'hombres' who bought her. When they refurbished the ship on a shoestring budget, the first thing they did was hoist out the engine, never to replace it. Their statement was clear: fossil fuels had no place in the future of shipping. Joe's Garage was and remains a hotchpotch of anarchists, hippies, artists, and sailors. No wonder the idea appealed to many of the regulars, me included, who spent countless hours imagining a better world.

But is cleaning up the shipping industry as easy as returning to sail and getting rid of engines? As a keen sailor and committed environmentalist, I wanted to find out. When researching the 'sail cargo movement,' embarking on a sailing cargo vessel is a must.

Activists and entrepreneurs like telling stories. Sail cargo companies, which operate small traditionally rigged vessels to transport cargo, are run by people who combine elements of both. They want everyone to believe in the future they envisage. Like most entrepreneurs, their most significant asset is the story they sell. Unlike most climate activists, they have hundreds of thousands of euros invested in companies, hoping to make a difference. It is important to understand their motivations, which is why I greatly appreciate the time people like Cornelius Bockermann of the *Avontuur*, Jorne Langelaan, Arjen van der Veen, and Andreas Lackner of the *Tres Hombres*, Danielle Doggett of the *Ceiba*, and many others have taken to speak to me. Rather than taking these well-meaning shipowners' stories at face value, I decided to sail along and research the complex environment they operate in, by also speaking to the crews choosing to live and work aboard for months on end. More than that, I became one of them.

In February 2020, I embarked on the *Avontuur* in Santa Cruz de Tenerife on the Canary Islands and sailed to Veracruz in Mexico, with stops in the French Antilles, Honduras, and Belize. On the way back to Germany, we called at Horta – the mid-Atlantic sailing capital in the Azores – and rounded the UK along the north, between the Shetlands and the Orkneys.

I never meant for my trip to be this long. I joined the ship in Tenerife intending to disembark some three weeks later on Marie Galante, a tiny island in the French Antilles just south of Guadeloupe. After that, I would make my way to Costa Rica's

Pacific coast to visit the Astillero Verde, the shipyard where Danielle Doggett, Lynx Guimond, and their team were building the *Ceiba*, a 46-metre 'zero-emission' sailing ship built to carry 250 tonnes. Needless to say, my 2020 did not unfold as planned.

In early March, as we were midway across the Atlantic Ocean, Cornelius Bockermann, owner of the *Avontuur*, sent our captain a message: the world as you know it no longer exists. As borders closed to curb the spread of COVID-19, my time at sea extended, port after port. Three weeks became five months. While I did not live through those initial months of fear, wonder, and frustration ashore, I much prefer my lockdown story of 2020 to the experience of almost anyone else I know. For one, when you're at sea, nothing changes all that much: leaving the ship is always impossible. Being denied shore leave was frustrating at times, but our everyday life at sea barely changed while the world back home was turned on its head. Our community of fifteen continued to function well, despite some inevitable tensions.

It helped a great deal that we did not have much of an internet connection on board – only the captain had direct contact with Cornelius through an expensive satellite link. Our knowledge of the virus's spread was both limited and intermittent. Not knowing when and where we'd be able to disembark bred frustration, but knowing that virtually all crews across the merchant marine shared our plight brought solace.

By the time we arrived in Hamburg, Germany on July 23, we had crossed the Atlantic Ocean twice, sailed the Caribbean Sea, the Gulf of Mexico, the Straits of Florida, and the North Sea. I lived aboard for 150 days while covering 13,957.5 nautical miles. During that time, I learned a great deal about myself, about sailing a cargo ship, and about the connections between the shipping industry and attempts to reverse the climate emergency.

On this journey, I encountered stevedores, shipping agents, customs officials, and ships' pilots. But I mostly saw my fellow fourteen crew members with whom I spent five months in close quarters aboard a century-old schooner. Between leaving the Canary Islands and arriving in the Azores, four months later, I did not once set foot ashore, even when we were in port to load cargo. We were, alongside hundreds of thousands of other seafarers, stuck on the commercial ships at the heart of modern trade flows.

The global 'crew change crisis' during the COVID-19 pandemic highlighted the perils seafarers face. They are essential workers in global supply chains, but their wellbeing is seldom noticed by the world's consumers. Six months after borders started closing all around the world in March 2020, more than four hundred thousand seafarers were still stuck at sea – beyond the ends of their labour contracts that can keep them at sea for up to eleven months – unable to go home.[7] A year on, in early 2021, when more and more people around the world started to receive COVID-19 vaccinations, this brought minimal improvement for seafarers. They come, in overwhelming numbers, from countries that lagged behind in their efforts to vaccinate. This slowness in the roll-out of life-saving protections not only poses a risk to crews' health and safety, but also to the global supply chains that rely on shipping.[8]

Shipping companies expected the global economy – and with it demand for shipping – to contract as a result of pandemic-induced lockdowns. While they cut capacity, that contraction didn't quite materialise. Stuck at home, people consumed more than ever. Demand for goods and thus cargo transport surged, driving freight rates to tremendous heights, creating a perfect storm of strained supply chains due to high demand, low capacity, and a crew change crisis that proved difficult to resolve.[9]

The pandemic did not create the racialised inequalities that characterise maritime labour. The colonial legacy of lascars, underpaid seafarers from the Indian subcontinent, endures in the use of 'flags of convenience' that allow shipping companies in rich countries to recruit seafarers from low-wage countries and pay them at rock-bottom rates. The legal fiction that the ocean somehow exists as separate from life ashore allows ships to fly flags of convenience, also known as flags issued by open shipping registries. Industry regulations allow shipowners to choose where in the world they register their ships, often leading them to opt for a handful of countries such as Panama, Liberia, and the Marshall Islands, which make it possible to skirt labour, safety, and environmental regulations that exist in rich countries.[10] While these flags of convenience allow the worst excesses in terms of labour conditions, life at sea is hard at the best of times. Seafarers can be at sea for up to eleven months, working long hours in difficult conditions, while away from home and family. Meanwhile, shipping companies recorded record profits throughout the pandemic.[11]

The historical exceptionalism of labour at sea has meant that the shipping industry is yet to accept eight-hour working days and forty-hour working weeks.[12] Seafarers' working hours are regulated under the Maritime Labour Convention. Ordinarily, they work seven days a week in shifts, creating working days that should be no longer than fourteen hours and working weeks no longer than seventy-two hours, with ten hours of rest a day and seventy-seven hours per week as a strict minimum.[13] During emergencies, these rules are suspended. For fear of being blacklisted by shipping companies or the manning agencies they use to hire crews, workers have little choice but to work longer hours if their captain or commanding officer tells them to.

Being stuck at sea introduced me to seafarers' labour and the complex patchwork of rules that governs their waking – and sleeping – hours. But my being aboard a German-flagged vessel with a mostly volunteer crew, all of us with European or American passports, made it difficult to transpose the experience to the industry as a whole. And yet, through conversations with my fellow crew, I learned why people volunteer their time and hard work to make sailing cargo vessels possible. If shipping companies won't even pay people fair wages, how would they turn the shipping industry, which transports 11 billion tonnes of goods a year, from a massive polluter to a zero-carbon model of sustainable transport?

In externalising the social and environmental cost of shipping to the high seas, the shipping industry mirrors the collective action problem that is the climate change to which it contributes. Every country wants to connect its economy across the oceans, but few feel responsible for the social and environmental impacts of shipping, or indeed climate change. This book recounts the personal odyssey that opened my eyes to the sheer complexity and scope of the challenges we face, three decades since the Rio de Janeiro Earth Summit in 1992. It also charts the journey the shipping industry must make to cut its carbon emissions. Will the industry make the urgently needed shift away from horrendously polluting, but frighteningly convenient, fossil fuels to emission-free propulsion through a mix of wind propulsion and 'zero-emission' e-fuels? And in doing so, will this meaningfully help mitigate the climate crisis that is rapidly unfolding? I hope so, but I'm not sure it will.

Decarbonising the propulsion of ships is enormously important and urgent, but narrowly focusing on this energy transition overlooks scores of other environmental challenges. When looking at the future of humanity beyond the immediate need to stop using fossil fuels, I wonder what kind of a shipping industry we need on a liveable planet. Or, to connect shipping to the broader economy it enables, what would make a socially and environmentally sustainable economy, which sits between the social foundation all humans need to live dignified lives and the environmental ceiling that sets the boundaries of what the planet can sustain?[14] This requires us to think not just until 2050, the time horizon for most climate goals today, but hundreds – if not thousands – of years into the future.

From this perspective, the mission of the *Avontuur* and other sailing cargo vessels remains as intriguing as ever: showing that a different kind of global economy is possible. Or is it?

1
Departure

Tenerife: carnival and rum

First impressions count.

I was nervous, hoping to put a good foot forward that night as I stood with Cornelius Bockermann, on a cold and windswept quay in the port of Santa Cruz de Tenerife. I was about to join the *Avontuur* on a transatlantic voyage, and I'd never spent such a long time at sea. The ship was, by that point, well behind schedule. She left her home port of Elsfleth, a small town on the river Weser, on January 17, with a week's delay. She'd had to wait for favourable winds to leave the Weser estuary near Bremen, where she was moored for the winter.

We were waiting for the *Avontuur* to dock, cold beers at the ready.

Coming alongside *port side to* with the wind pushing her bow away from the quay proved difficult. After two failed attempts, the lines were on.

I felt like a school kid joining a class mid-year. How would I find my place? Whom would I get along with? I knew from experience just how quickly sailing crews establish camaraderie, routines, and hierarchies. This crew had been aboard together for more than a month, without a moment ashore.

They had braved a storm in the North Sea, sat out rough weather in the Bay of Biscay anchored off Douarnenez in Brittany, and drifted around in the windless sea off the Portuguese coast. Before that, they had spent more than a week together in Elsfleth, readying the ship for her voyage.

There I was, next to the big boss, waiting to get my first glimpse of the ship's crew. The crew, however, seemed more interested in cold beer and carnival celebrations than in meeting me. Nothing makes the heart grow fonder of beer-fuelled celebrations than working the sails in long watches on a dry ship.

And me? I stood sipping my alcohol-free beer.

No matter how interesting we researchers think we are, we don't have a reputation for being fun. Benji told me, months later, that I looked like a stiff and boring academic. Being a teetotaller only made things worse. I'd need a good excuse for joining the crew in sunny Tenerife, rather than struggling through the North Sea in a January storm like everybody else.

In short, I had some explaining to do to my shipmates.

They'd all joined the *Avontuur* in the early days of January at the Elsfleth shipyard. She was in the water, but nowhere near ready for a six-month Atlantic round trip. Part of the privilege of sailing a ship like the *Avontuur* is the opportunity to learn how to maintain a sailing vessel. Entropy does not rest, and neither can the crew.

Cornelius intended to sail the *Avontuur* on her fifth voyage under his ownership. Despite being the owner of the ship and a fully qualified master mariner, he does not skipper all of her journeys. Sailing a ship is a full-time job. But so is running a shipping company ashore. And for all the advances in digital communications, running a shipping company on board while at sea simply does not work.

Paul Wahlen, a former owner of the *Avontuur*, also knew this. He ran the vessel as a tramp-shipping company – that is, sailing without fixed routes or ports of call by responding to ad-hoc transportation needs. But doing that in the last quarter of the twentieth century was even more difficult than it is today. Wahlen, much like Brad Ives who sailed the *Kwai* between Pacific Islands until the end of 2020, sailed more for lifestyle reasons than to make a point about the lack of environmental sustainability of the shipping industry.

I was looking forward to sailing with Cornelius, as I hoped to get to know him better. He did, after all, plunge his entire savings and years of his life into the project, but days after the crew had gathered in Elsfleth, Cornelius fractured several bones in his hand. So, the *Berufsgenossenschaft Verkehrswirtschaft* – the Trade Association for the Transport Industry – declared him unfit for work. Within days he'd have to find another master mariner with a commercial licence, endorsed for Germany.

Then the First Officer had to pull out of the voyage for lack of the right paperwork. Next, the Cook withdrew for personal reasons. And while Cornelius scrambled to find replacements at short notice, departure was pushed back until storm Gerlinde, which plagued northern Germany January 13–14, blew over.

Meanwhile, I was still in Australia. I did not fly to Europe until January 21. I was keen to leave Melbourne, where the Black Summer had caused terrible bushfires that burned an enormous area, estimated to be between twenty-four and thirty-three million hectares, which killed or displaced three billion animals.[1] The air quality around much of the country was so bad people opted to stay indoors.[2]

I planned to join the ship upon her arrival in Tenerife, scheduled for February 8. As I followed the *Avontuur* on *Marine Traffic*, delays were adding up. Meanwhile, I found myself in Amsterdam, reading and writing in cafes and libraries, staying with friends, and speaking to Jorne Langelaan in Alkmaar, who had left the *Tres Hombres* to focus on his new venture the *EcoClipper*, an hour north by train. It was cold and rainy, and I was getting bored. So, rather than waiting for Cornelius to give me the final joining date, I flew to Lanzarote where I rented a rickety old camper van and toured the island. I spent my days swimming, hiking, and eating on the rugged volcanic island. As the arrival of the *Avontuur* neared, I embarked on the *Volcán de Tamadaba*, a slow old Naviera Armas ferry to Gran Canaria and then onwards to Santa Cruz on the *Volcán de Tagoro*, a fast catamaran ferry, built by INCAT in Tasmania.

'The engine is there for emergencies,' Cornelius told me some days later over a coffee at the Tattoo Art Café, opposite the Teatro Guimerá in Santa Cruz, 'not for the comfort and convenience of the crew.' He told me this because the *Avontuur*'s master, whom I'll call Captain, had turned on the engine along the Portuguese coast. This information went public as the AIS, the maritime Automatic Identification System that transfers a ship's information via VHF, includes information on the type of propulsion a ship uses.

'This is not the spirit of the company,' Cornelius told Captain via email. *Marine Traffic*, a popular ship-tracking website, allows anyone to see the position and movement of any ship equipped with AIS. As long as the ship is close enough to shore, this information is updated in real time. Propulsion data fuels the arguments of sail's detractors. 'How can anyone seriously propose returning to sail?' they might ask. 'Even at such a small scale, it can't be done.'

It is difficult enough to convince shipping magnates that sailing is more than just fun, but also a commercially viable propulsion technology in the twenty-first century. Online exposés of sailing cargo ships motor-sailing for lack of wind does not help.

'There will be more delays,' Cornelius reminded me, or perhaps mostly himself. 'Being seasick and waiting for the wind is part of what everyone willingly signed up for.' And wait for the wind we did, as the *Avontuur* was stuck right in the middle of a calm, directly north of Tenerife's headland. He seemed tense and impatient. Hardly surprising given the money and time he's invested in Timbercoast, his shipping company.

I did not mind waiting at all. As the days passed and the *Avontuur* inched closer, I met Cornelius' brother who lives on Tenerife, and who had agreed to help find organic produce to re-provision the ship. Meanwhile, I had discovered the beautifully modern municipal library, the excellent restaurant of the *Museo de la Naturaleza y Arqueología*, and the quaint university town of La Laguna, up in the hills a mere tram ride away. I could easily see myself living on the island and was in no rush to leave.

'I would have turned the engine on now,' Cornelius said, contradicting his earlier commitment to fully engine-free sailing. 'At four knots over calm water, she uses some six litres of diesel per hour. Motor sailing at such a speed for a short period would ensure she didn't end up on the wrong side of the island, which could easily happen with the slightest puff of easterly wind.'

Cornelius, a master mariner, entrepreneur, and former boss of many, is used to getting his way. Working with another captain as master aboard his *Avontuur* complicates that. A ship can have many captains (as this is a personal title), but only one master at a time. Only the captain in charge of the vessel is its master, superseding the owner's authority over the vessel when at sea. Irrespective of their maritime rank, whatever the owner says is moot unless they are at sea as master.

Cornelius agonised about telling Captain to turn on the engine – it would mean swallowing his pride and principles. He could not be seen to backtrack on his criticism. Nor could he publicly carry the blame for undermining Timbercoast's principles and objectives. A lot was at stake.

Captain, I would soon learn, was equally stubborn in his refusal to ask Cornelius' permission to fire up the engine.

The sail cargo pioneers of Fairtransport, who run the *Tres Hombres* on a shoestring, don't have this problem. For lack of an engine, they have to simply wait for the wind. But that's not the only thing that sets them apart.

The *Avontuur*, with her German flag, expensive ECDIS navigational equipment, water maker, and relative comfort – compared to the *Tres Hombres* at least – plays fully by the book. Doing so costs an awful lot. The *Tres Hombres* also plays by the book, but it's a wholly different one. It is exempt from registering under the IMO's Ship Identification Number Scheme, as it is a wooden ship without 'mechanical means of propulsion.'[3] Not having an IMO number, despite being a commercial vessel flying the Vanuatu flag, means that the *Tres Hombres* can avoid the scrutiny of Port State Control during cargo operations.

At face value, their mission is the same – decarbonising the shipping industry. But these companies are run by people with vastly different careers and life experience. Cornelius' background running maritime businesses in Nigeria gives him an excellent understanding of how the conventional industry works. But the shift from subcontracting for the oil industry to running a small and independent shipping company is difficult. The costs add up, day after day, and carrying a few thousand bags of coffee simply doesn't pay like the oil industry does. Revenues are tiny, but running costs are high.

The founders of Fairtransport have plenty of sea time too. But unlike Cornelius, they know how to operate on a shoestring. They run an organisation that is well integrated into the Dutch gift and mutual-aid squatter and activist networks.

In principle, these different initiatives make up the Sail Cargo Alliance, a network of companies. But in practice, sail cargo initiatives are stronger on creating a semblance of collaboration than on genuinely working together. The *Avontuur*'s Tenerife stop would soon illustrate that.

The day after docking in Santa Cruz, I boarded the *Avontuur*. Finding a berth involved a complex game of musical chairs, as those stepping off had not yet vacated their bunks. Eventually, I found myself in the smallest bunk of the fo'c's'le, nearest to the heads. It was probably the least salubrious bit of real estate aboard. But it was to become my own domain.

With curtains separating the coffin-sized bunk from the other nine occupants of the sixteen-square-metre space, privacy was an illusion. Two simple rules strengthened that illusion: entering a bunk other than yours was *verboten* and closed curtains meant 'do not disturb.'

With only a thin metal sheet between my head and the hand-pumped head (boat-speak for toilet), not being disturbed meant little more than 'no small-talk.' The forward head, located in the fo'c's'le in the front of the ship, whom we'd later call Ernest because he had a life of his own, would gurgle and burp unexpectedly. When lucky, he'd only let out a waft of foul-smelling gas from the black water tank; when unlucky, he'd spit out the contents of the plumbing pipes. Flushing required dragging a bucket of seawater down to the head. With, of course, the request to kindly leave a full bucket for the next visitor. As Ernest was the only forward head, serving ten people, there would be a constant symphony of pumping and gurgling. Day and night. The aft head, at the rear of the ship, was located above Captain's cabin, in the deckhouse next to the galley and the chart room.

Aboard ships, one doesn't use front and rear nor left and right as these are relative directions, depending on the position of the person using them. To avoid confusion, the 'front' of the ship as a physical place is the 'bow,' the front as a direction is 'forward' or 'fore' while the 'rear' of the ship is the 'stern' or 'aft.' Left is 'port,' whereas right is 'starboard.' This means that aft starboard is always in the same place, irrespective of where one stands.

The galley, fiefdom of our Cook, somewhere between Canada and Horta, June 18, 2020

Between painting the deck, stowing provisions, and getting inebriated ashore, the crew were busy. Meanwhile, we awaited a barrel of rum that never arrived.

Cornelius had put in an order for a barrel of rum, but the Aldea distillery (which also supplies the *Tres Hombres*) told him there was no stock. Cornelius learned that others were able to put in orders, so he asked a friend on Gran Canaria to buy a barrel and ferry it across to Tenerife. But once the distillery got wind of this, they offered bottle-proof (rather than the stronger barrel-proof) rum for a price that would make it hard to turn a profit. Some blamed Cornelius for following a bit too closely in the footsteps of the *Tres Hombres* by trading the same goods on the same route, while others lamented the lack of true collaboration through the Sail Cargo Alliance.

Meanwhile, carnival put a hold on port operations and retailers, forcing us to wait until celebrations were over before setting sail.

No one minded the parties. But when the Calima, Saharan dust carried across the ocean, struck the Canary Islands, it left us with horrendous air quality, which brought back my memories of the Australian bushfires I'd left behind the month before.

While others were out celebrating, I volunteered to take evening harbour watch. It gave me some time to talk Captain, the master of the ship, through my research and objectives. I found him very open to both my project and approach. Even so, it was difficult to get past his stern and rigid guard. He made it clear that he preferred the long-form and in-depth engagement of an academic over self-aggrandising bloggers, whom he seemed to despise, along with any and all social media.

Irrespective of Cornelius' views as shipowner, I'd have to ensure the master of the ship would be happy for me to be there as a participant observer.

The next day, after the First Officer arrived, I formally told the crew my reason for being aboard: research. Several veteran crew had encountered journalists on previous voyages who had violated their trust and personal space. They told me they wanted to be sure they could trust me and valued the project I was working on. Everyone consented, some anonymously and some by name.

I will refer to the professional crew mostly through their roles: Captain (Michael), First Officer, Second Officer (Pol), Bo'sun, and Cook (Giulia). As there were two deckhands on board, I will refer to one of them as Little Captain, for reasons that will soon become obvious. The other one is Joni. Meanwhile, some of my shipmates get nicknames, based on how they were known aboard. Some said, though, that they'd be happy to be cited by name. Jennifer, Benji, Athena, Martin, and Peggy go by their real names only. Mia and Pinkie are nicknames.

There are no shortcuts to sailing a ship like the *Avontuur*. She needs constant care and attention. Every watch, every hour, every last moment, the crew needs to ensure that the ship and its cargoes remain safe, meaning that a crew always shares an existential concern for the state and safety of a vessel. The ship is not just a means of moving goods, it is also one's home. And it needs constant attention and work to stay afloat and protect us against the elements.

As we prepared to set sail, Cook told Cornelius with enthusiasm that 'Captain can't leave in Marie Galante' as was planned, or 'all the crew will quit.' This resounding vote of confidence and testament to the popularity of the Captain was a good sign. It's not easy as a ship's captain to balance being in charge and being liked. As ship's master, any captain needs to retain authority above all else. Being popular is a lesser concern. Historically, captains who lived in their cabin in the aft of the ship, where the officers also resided, preferred to maintain ample distance from seafarers 'before the mast,' who lived in the ship's forecastle or fo'c's'le for short, and officers were expected to do the same. It was a matter of class and status – which both Richard Henry Dana Jr and Eric Newby documented very well, from their perspectives as upper- and middle-class sailors before the mast. Unlike Dana and Newby's ships, the *Avontuur* had no dirt-poor seafarers before the mast, nor were our captain or officers upper-class gentry.[4] Today, a captain's authority, especially on a large sailing ship, is more a matter of law and respect for training and expertise than a difference in social class or status – even if the globalisation of the maritime workforce has created racialised hierarchies that echo earlier class-based hierarchies.

With carnival behind us, we set sail on Thursday February 27. As the last line was cast off, we left the permanent abundance of everything and with it, carnival's hedonistic excess in our wake. Cornelius looked tired and tense, but also happy to see us set sail. He'd meet all of us again in Marie Galante, where he'd take over from Captain, by which time Cook would sing to a very different tune.

No one had any idea what the voyage ahead held in store for us.

Marie

On the morning of February 28, within twenty-four hours of leaving port, my fellow shipmate Athena woke me at 07:00 in the morning. Breakfast would be ready by 07:30 and at 08:00 our watch would relieve hers, who had been on deck since four. Waking up the oncoming watch is a key duty of the standing watch. If you forget to do so, your own time on deck could extend beyond the scheduled four-hour slot. As I left the fo'c's'le to head aft, towards the galley, Cook asked me to take a few milk cartons with me, as she could not carry all those she'd taken out of the storage bilge herself.

As I emerged from the steep companionway steps at 07:40, with six cartons of *Frische Bio Alpenmilch* – fresh organic alpine milk, Bo'sun told me to drop everything and help take down all sails. Right away.

I didn't get what was going on. We seemed to be making good speed in light winds on a calm sea. Why the rush? Why take down the sails at all? Under normal circumstances, the oncoming morning watch has time to stretch their legs, wake up with coffee and breakfast before taking to the deck.

At 07:36 that morning, the lookout had reported a 'small boat with people, whistle sounds.' Within ten minutes, we had taken down sails and started the main engine. By 08:00, we had come alongside the *Marie*, with her five women and eleven men aboard.

We were close enough to hear them, though not quite near enough to do more than listen.

They told us they'd been at sea for ten days. Ten days of utter disorientation in the vast blue ocean, devoid of shelter of any kind. *Marie* was an open wooden fishing vessel, painted blue like most along the Moroccan coast. It was some five metres long and two wide. Its bilges sloshed with a mix of seawater, petrol, and whatever other liquids never made it overboard. We might have been close to the Mauritanian coast, but winter nights at sea – even at these latitudes – get very cold. They had run out of fuel, water, and food. Most of all, they were scared.

The night before, during our first Red Watch from eight to midnight, Captain peered at the chart as I stood next to him in the chart room, a tiny alcove port of the aft companionway. He explained how we'd sail south until we got close to Cape Verde, after which we would turn west to make a great circle to the Antilles.[5] For several days, the route would take us quite close to the West African coast. Rather casually, Captain mentioned that ever more refugees were attempting the sea crossing from Western Sahara, where the desert drops into the ocean, to the jaggedly volcanic Canary Islands. Often in unseaworthy boats like the *Marie*.

The shortest possible route they could take was from the Moroccan port of Tarfaya, just west of the Khnifiss National Park, to Gran Tarajal on the east coast of Fuerteventura. At only fifty nautical miles (some ninety kilometres), this is far shorter than the 160-nautical mile Mediterranean route from the Libyan capital Tripoli to Lampedusa. Leaving from Laâyoune (meaning *The Springs*) in the Western Sahara seems more attractive, as the former Spanish colony (from 1884 until 1976, shortly after Franco's death) is occupied by Morocco, despite being contested by President Brahim Ghali of the Sahrawi Arab Democratic Republic, which controls some 20 percent of the territory. As a result of the on-going contested occupation of the territory, and since the 1975–1991 Western Sahara War with occupying Morocco, many indigenous Sahrawi live in limbo across the Algerian border in Tindouf. This porous territory is an increasingly popular point of departure for refugees from across West Africa.

One of the risks of embarking in this region is drifting along with the south-bound Canary Current. This risk is real, as refugee vessels seldom have enough petrol to complete their planned journeys. Absent navigational equipment or experience at sea, any such voyage is a lottery. It can result in a new life in Europe. Or death.

Captain called the Spanish Maritime Rescue Coordination Centre (MRCC) on Channel 16, the international distress frequency, with no response. He then called Cornelius on our Iridium satellite phone who, without hesitation, supported Captain's decision to take all of the *Marie*'s occupants on board and bring them to the nearest port. Cornelius would organise support ashore with the Spanish MRCC.

Before embarking the *Avontuur* in Santa Cruz, I watched *The Left-to-Die-Boat*, a harrowing documentary by *Forensic Oceanography*'s Charles Heller and Lorenzo Pezzani (2012). It traces the voyage of a rubber boat carrying refugees from the Libyan capital Tripoli intending to reach Lampedusa. These refugees had a satellite phone aboard, on which they were able to call Father Zerai, an Eritrean priest in Rome, who is often a first point of contact for East African refugees. He notified the Italian MRCC, who received their coordinates from the satellite phone provider. Despite being sighted by a French aircraft that passed on the vessel's coordinates as well as two military helicopters, no one rescued the passengers. They were left to die.

Of the seventy-two people aboard when leaving Libya, only eleven were alive when the vessel made landfall in Zlitan, back in Libya. Two more died shortly after.[6]

The documentary that tells this story is a visual representation of the Mediterranean Sea. A chart. The story is pieced together with data points of distress calls, sightings, and coordinates known by Thuraya, the satellite phone provider the passengers used. The film does not show any beaches or ships, nor seascapes or talking heads. It depicts a chart on which the story slowly unfolds to its horrifying, but entirely avoidable, conclusion.

With this story in mind, I had little hope for a good outcome as I watched the *Marie* and her fearful occupants bobbing around on the open ocean. Would the Spanish coastguard send a Search and Rescue team? Or would they send us away when reaching shore?

By 08:30, we had *Marie* in tow, using a longline. We were slowly headed in the general direction of Gran Canaria, the island closest to us, before the MRCC Las Palmas advised us to set course for Puerto de Arguineguín, on the southern coast of Gran Canaria.

The unanimous desire aboard the *Avontuur* was to help the people in need. We did not have much choice, as it is a legal requirement under maritime law to help anyone in distress. But recent legal prosecutions of humanitarian organisations heading out to sea with the express purpose of saving refugees from drowning has made helping people at sea far more challenging.[7] At the same time, merchant and coastguard vessels regularly ignore distress calls, as heeding them causes delays, which cost shipping companies time and money and can spark political controversies for coastguards.[8]

By 09:30, all sixteen people from the *Marie* were aboard the *Avontuur*. We offered them water and food. They sat and lay on the cargo hold hatch, huddled in our woollen blankets. Despite the warm morning sun, they craved shelter and comfort.

'I had no idea the sea was so big,' one of them told us. As they drank, ate, and washed themselves, we listened. Absent showers and plentiful fresh water we put up a makeshift tarpaulin shower cabin-cum-bathroom. (Us crew ordinarily washed ourselves with buckets of seawater, rinsing off the salt with a tiny splash of precious fresh water.)

These people and their stories sounded familiar. They reminded me of the many West Africans I'd met when living and travelling in the region. Their stories of poverty and despair matched the ones I'd heard time and again. As we got to know each other, ever so slightly, Captain tended to their wounds.

It took another two hours before the *SAR Salvamar Menkalinan*, a Spanish Search and Rescue vessel, met us in the open ocean. They asked us to set the *Marie* adrift, as they'd bring the sixteen refugees safely ashore. We let the boat in which they'd spent ten long and

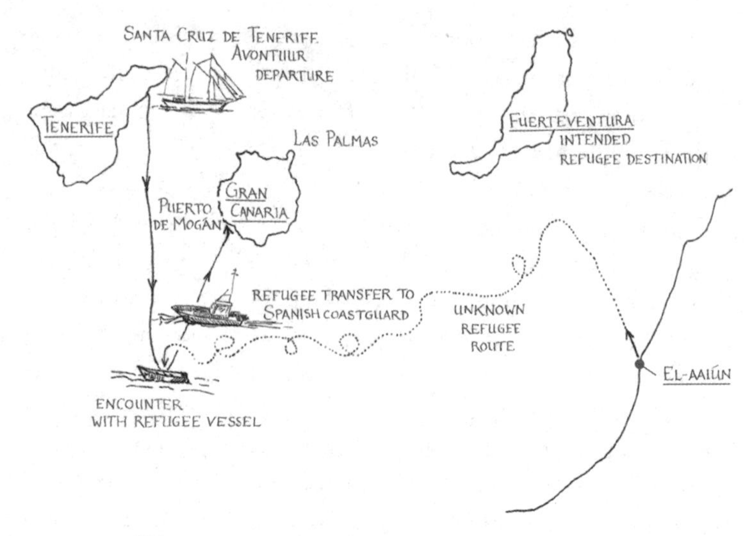

Where we encountered the *Marie*, February 29, 2020

horrible days and nights sink. It would pose too big a risk to keep it near when transferring the sixteen people to the other vessel.

They had survived. Even so, I feared for them. I dreaded their encounters with the European Union's 'immigration detention centres.'

I wish I could have shared their hopes for a new start on European shores. But, for many of them, I feared the struggle of arriving while coping with unspeakable trauma would stay with them for the rest of their lives.

'Look people in the eyes to see if they're good people,' our Cook, who had previously worked with refugees on the Greek islands, told one of the young women we had rescued. 'Learn Spanish, learn the language,' Cook continued.

'All will be well,' responded the young woman, 'God willing.'

'Si dieu veut, yes, but also if *you* want it,' Cook said as parting words, moments before the frightened refugee stepped aboard the *Salvamar Menkalinan*, where the others were huddled in emergency blankets on the ship's bow. 'Trust Allah, but trust yourself, too.'

The Middle Passage

Puerto de Mogán is a small, quiet town on the southern coast of Gran Canaria, which was not particularly keen on our arrival. Captain and Cornelius decided the *Avontuur* would head into port there after the sixteen refugees boarded the *Salvamar Menkalinan*. We needed to decompress after the experience and the ship was to be disinfected. Not because of COVID-19, but out of concern the sixteen people we'd encountered had brought some other communicable disease on board.

Little suggested a pandemic was in the making.

No other port allowed us entry. The news of the refugees had travelled fast. And our presence might have blemished the picture-perfect holiday destination that is Gran Canaria. Perhaps news arrived more slowly in sleepy Puerto de Mogán. Still, by the time we arrived, the locals seemed to know. While they did not turn us away, they were not very welcoming either. The harbourmaster did not want us alongside in port. The rocks protruding from the seabed would be a risk, they said. Though they could not tell us what the exact clearance was. The harbourmaster sent us to anchor outside the breakwater in the bay, from where we'd ferry ourselves in and out of town on Bob, our trusted dinghy.

The place reminded me of Salcombe, a quaint town in Devon, which I'd called at whilst sailing on the *Moosk* the year before. Visitors seem to do little but wander, eat, and drink on the sanitised waterfront. Behind the façade of leisurely consumption, a mere few streets from shore, Puerto de Mogán showed its true face: a bleak and little-loved backwater, reminiscent of just about any impoverished town. The rocky shoreline doubled as a rubbish tip and the local housing appeared grey and unwelcoming. The town and its visitors seemed to be held captive by the scraggy surrounding mountains.

British and German tourists loitered on the promenade, gorging on fried food and cheap alcohol. Everything on offer, from

The beach at Puerto de Mogán, Gran Canaria, February 29, 2020

Chinese-made plastic sandals to fresh water, seemed to be either imported or naturally scarce.

After our chance encounter with sixteen human beings desperate enough for a better future to embark on a tiny fishing boat, our mission suddenly seemed pointless. We hadn't even left Europe, and I was already questioning the purpose of our voyage. Why bother crossing the Atlantic under sail for coffee and chocolate as other people risk their lives to reach Europe? Did it make any sense to spend so much time, effort, and money on transporting a tiny amount of luxury cargo across the ocean? Did the symbolic act of sailing cargo justify the effort?

We were to cross the Atlantic and bring the same goods as the colonial triangular trade had brought to Europe: coffee, chocolate, and fermented and distilled sugar cane. Had nothing changed since the Atlantic had served to transport enslaved human beings as units of labour in a brutal economy that had only one aim: enriching white Europeans?

The Middle Passage was the second leg of the triangular trade route that made the modern world.[9] As a Eurocentric concept, the triangular trade involved three legs. The Outward Passage from Europe to West and Central Africa served to transport manufactured goods such as textiles, tools, and weapons. These were sold as merchandise and used to restock colonial settlements. The Middle Passage transported captured human beings from Africa to slave plantations across the Americas and the Caribbean. Ships departed ports along the entire western coast of the continent, roughly in between what we today call Senegal and Angola. The Inward Passage brought the spoils of the new world, such as tobacco, cotton, sugar, and of course coffee and cacao to Europe.

On the Outward and Middle Passages, our hold was empty. Today, only the Inward Passage offers a lucrative option for sailed goods.

'The slave trade,' Liam Campling and Alejandro Colás argue in *Capitalism and the Sea*, 'integrated the merchandise of captive Africans into the emerging world market, thereby contributing to the process of capitalist development in Europe.'[10] Ian Baucom's *Specters of the Atlantic*, recounting the *Zong*'s voyage in 1781, shows that the roots of 'modern' capitalism lie in the ways in which the slave trade was financed, insured, and its spoils shared by investors, more so than by merely trading captured human beings. This forces us to rethink the legacy of the Atlantic slave trade as not only deeply inhumane, but also as underlying the logic of the economic arrangement under which we live today.[11]

Our encounter with sixteen human beings taking a huge risk in search of a better future illustrated the legacy of the

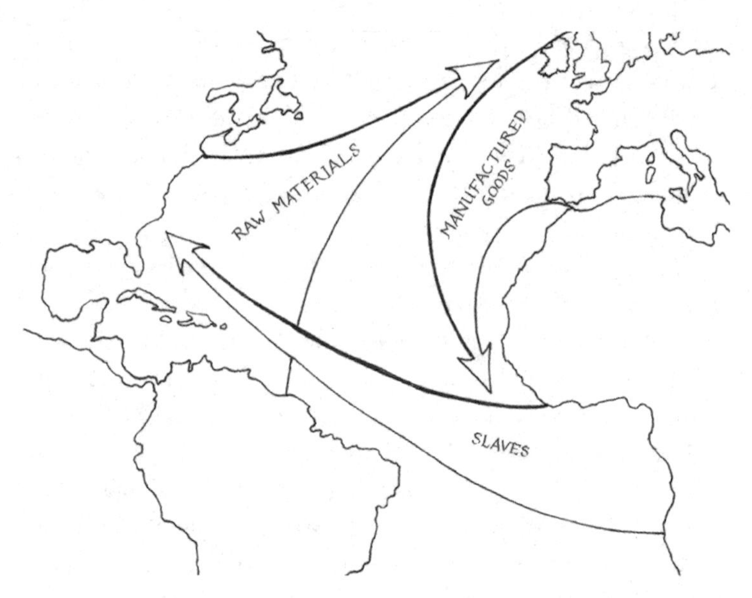

The triangular trade during colonial times

triangular trade. Without the damage done by colonial exploitation and slavery they might not have gone to sea. Without the neo-colonial exploitation of their 'independent' economies, or the climate change inflicted on them, they might have been able to build a better life at home.

Yet here we were, on a Spanish outpost in the Atlantic, just off the Moroccan coast, sailing an empty ship to the Caribbean, as if the Middle Passage never happened.

Slave ships, which once sailed from places like Île Gorée off the coast of Dakar in present-day Senegal and Elmina Castle in what is now Ghana, no longer cross the Atlantic. But the legacy of this trade persists.

The trade winds continue to propel sailing ships like the *Avontuur* across oceans. The north-easterlies that pushed us to West Africa before the easterlies took us across the ocean remain the same. The cargo we were to ship was still the same as when tall ships worked the trade winds in colonial times. Those ships, on their triangular route, ferried human beings as slaves across the ocean where their forced and stolen labour made coffee, cacao, and sugar plantations a lucrative enterprise. Slavery is now illegal but remains a global problem – as the use of child labour in cacao production shows.[12]

Before we set sail, I enthusiastically mentioned encountering a display of eco-feminist books at the *Biblioteca Municipal Adrián Alemán De Armas*, a public library in Tenerife, to Captain and Jennifer. Captain dismissed the value of any such works. 'Why would you stress the separation of men and women? We are all equal and dividing us like that does not do any good.'

Jennifer, an Irish filmmaker who had joined the ship as a voyage of experiential research for the scenario she was working on, rolled her eyes. Captain's simplification of gender and feminism overlooked precisely the point that eco-feminists make. Gender, they argue, plays a big role in the vastly unequal causes and impacts of climate change. In modernity's 'master mentality,' humanity equals the reason of men, in contrast to the presumed 'emotional and bodily femininity of nature'.[13] This tension is more complex and nuanced in our lived experiences, but the overlap between environmental and sexist hierarchies is precisely what eco-feminists expose and challenge.[14]

Captain went on to stress that he never had any issues between men and women on board. The *Avontuur* did indeed seem to do rather well in avoiding any overt gender disparities. With six women and nine men on board, our crew had a forty/sixty split. Considering the historically male environment that is the merchant marine, that was an improvement. Though the only paid female crew member was the cook.

'If there is a climate change problem, it is in large part a justice problem,' Mary Robinson, the UN Special Envoy on Climate Change stresses. 'Our continued existence on this shared planet demands that we agree to a fairer way of sharing out the burdens and the benefits of life on earth, and that in the choices we make, we remember the rights of both today's poor and tomorrow's children.'[15]

But how will we accomplish this sharing?

'Unless our culture goes through some sort of fundamental shift in its governing values, how do we honestly think we will "adapt" to the people made homeless and jobless by increasingly intense and frequent natural disasters?' asks Naomi Klein. 'How will we treat the climate refugees who arrive on our shores in leaky boats? How will we cope as fresh water and food become ever more scarce?'[16]

These difficult questions about planetary justice cannot be separated from the climate crisis we face. This is a recurrent theme in postcolonial scholarship. 'There is an explicit kinship between plantation slavery, colonial predation, and contemporary forms of resource extraction and appropriation,' argues the postcolonial theorist Achille Mbembe. 'In each of these instances, there is a constitutive denial of the fact that we, the humans, coevolve with the biosphere, depend on it, are defined with and through it and owe each other a debt of responsibility and care.'[17]

It is as if humans have lived in a 'fantasy whereby we make our existence viable by seeking homeliness through aggression and domination.'[18] First through the exploitation and dispossession of the racialised 'other,' now through the rapacious plunder of the fossilised remains of plants that withered millions of years ago.

In much the same way that the slave trade pillaged Africa, what Naomi Klein identifies as a continuing history of 'extractivist' economics is now harming the planet. We now live in the Anthropocene, an age in which human activity is changing the environment.[19] It is not, as some argue, a uniformly anthropocentric development, however. Human activities are not equally responsible for Earth's transformation, nor does it affect people in the same ways.[20]

Those who have benefited least from the human domination of the planet will likely suffer the effects of climate change most.

And millions of them, perhaps as many as two hundred million by the year 2100, may become climate refugees.[21]

'Our economic system and our planetary system are now at war,' argues Naomi Klein.[22] But how do we intend to deal with the refugees fleeing this planetary war? The challenge of accommodating climate refugees is only just starting. Whatever their exact number, it will far exceed the eighty-four million refugees in the world – and even that enormous number dates from before the Russian invasion of Ukraine (itself arguably a climate war).[23]

Will we only welcome refugees *after* disasters? Or will we accept that some may seek refuge before their house is flooded by rain or a rising sea? Before their house goes up in flames or gets flattened by a storm?

An underlying issue is that citizens have been reduced to consumers. We are no longer political actors. We are solely consumers of a fossil-fuelled and globalised economy.[24] We have lost the ability or even the desire to halt climate change, to walk off the job.

For the job is not us. The job is the pipeline that keeps pumping. Hence Andreas Malm's proposal that we move beyond peaceful protest and sabotage fossil-fuel infrastructure; that we start blowing up pipelines.[25] Or, as I'll suggest later, stop shipping fossil fuels.

'Racism,' Ghassan Hage writes, 'is an environmental threat because it reinforces and reproduces the dominance of the basic social structures that are behind the generation of the environmental crisis – which are the structures behind its own generation.'[26] It is the violation of respect for other human beings and nature that has allowed a rapaciously dominant group to take the reins of the planet, first through colonisation, then through empire, now through economic domination. He suggests that 'one cannot be anti-racist without being an ecologist today, and vice versa.'[27] The 'domestication' of the racialised other follows the same logic as the subjugation of nature to the will of humans; or rather of rich and powerful white men.

The desperation of the suffering souls of the *Marie*, whom we encountered adrift in an unforgiving ocean, is powerful testimony to the enormous damage that continues to be done to human societies and our planetary environment. Not to mention those unfortunate enough not to survive their journey.

The preamble to the United Nations Paris Agreement lists a range of assumptions and ideas that the political compromise builds on. This is part of the format in which the United Nations presents its texts. It uses this approach to connect new frameworks and agreements to its existing web of ideas. But sometimes this section is also used to relegate crucial issues to the margins of the document at hand.

> *Noting* the importance of ensuring the integrity of all ecosystems, including oceans, and the protection of biodiversity, recognized by some cultures as Mother Earth, and noting the importance for some of the concept of 'climate justice', when taking action to address climate change.[28]

It is a clear message: climate justice is important *for some*. There is no moral or legal basis for climate justice in the Paris Agreement – even though there is some legal-technical provision, such as the principle of *Common but Differentiated Responsibilities and Respective Capabilities*, about which more later on.[29] Rather than being a great equaliser, the climate crisis is doing the opposite. And the warning that those least responsible for climate change will bear the brunt, the warning that people in the 'Global South' will suffer more, seems to encourage inaction in rich countries rather than compassion and decisive action.[30] Though 'the rich have much to lose; the poor do not.'[31]

It is not by virtue of merit that some are born rich and others poor, or that some live in safer environments and others in more polluted ones. Our citizenship is, as Ayelet Shachar argues, a 'birthright lottery,' over which we have no control whatsoever.[32]

Shipping made global trade, colonisation, and the spread of capitalist empires around the world possible. And it is the fuel-dependent globalised economy that is driving the climate emergency that sail cargo initiatives are trying to halt. Whether the passengers of the *Marie* were fleeing postcolonial poverty or climate change scarcely matters. The economic successes of colonisers relied on maritime transport that directly or indirectly led to the overwhelming problems that these sixteen humans were trying to escape. Simply decarbonising ships' means of propulsion doesn't solve any of these underlying issues. But can an initiative as symbolic as a sailing cargo ship do anything to turn the tide on both climate change and the history of colonial exploitation, with the deeply dehumanising slave trade at its core?

It was when 'free' labour – that is, stolen labour – no longer existed, after the outlawing of slavery, that the dominant human societies shifted to another form of 'free' labour: burning fossilised plants that had absorbed sunlight millions of years ago. The fossilisation of these plants was like an enormous carbon-capture scheme spanning millions of years. It drew vast amounts of carbon from the Earth's atmosphere. This, in turn, created the chemical composition of the atmosphere that resulted in the stable climate which allowed humans to flourish on the planet.[33]

Even so, our encounter with the people of the *Marie* did not seem to generate much debate throughout the remaining voyage. Not about the history or the contemporary reasons why some people flee their homes.

Understanding maritime transport is crucial to understanding global history. The globalised economy would not exist, for better or worse, without shipping, but historical accounts often present the sea as a seemingly frictionless and liminal space, where the struggles of the land are largely suspended. Most history books treat shipping as something that just happens; as something that does not have a complex history of its own. Crossing the watery bits of history was never a minor detail in the making of a globally connected world; sailing made the modern world, including slavery and its legacies.[34] Looking beyond the blue horizon, as Brian Fagan suggests, required imagination, courage, and skill.[35]

And yet, at 06:00 on March 1, we set sail again. To pick up coffee, cacao, and rum.

It was only later, after ten days at sea, that I started understanding just how long these people had been lost at sea aboard *Marie*, a tiny open boat with an outboard engine. They had none of the homely comforts, navigational tools, food supplies, water stores, seafaring skill, and safety equipment we counted on.

My first ten days at sea felt like months. Theirs must have felt like a miserable lifetime.

Living before the mast

As we sailed out of Santa Cruz de Tenerife, before we encountered the *Marie* and her sixteen passengers, sailing was smooth. The breeze was crisp and fresh, the winter sun bright. At seven knots, we were sailing 'fast enough,' Captain said. As a sailor used to yacht racing, I had to get used to this attitude. There is no 'fast enough' in a race – unless you're in it to see someone else take the trophy. I personally never cared about winning any of these races, as I was in it for a good time on the water. Though engaging in racing does instil a habit of pushing a vessel to the limit.

Our weekend club races around Port Philip, from the Sandringham Yacht Club in Melbourne, on the *Windspeed* with her nonagenarian skipper, rarely lasted more than a few hours. On the *Avontuur* we had an ocean to cross. With that timeframe in mind, the *Avontuur*'s Captain clearly favoured safety and endurance over speed. Months later, when confronted by far more unstable weather and sudden squalls in the Gulf of Mexico, we would find out why.

As we got going, I offered to take over the helm.

'Do you know how to helm?' Captain asked.

'I'll try,' I responded.

'We don't try,' he responded. 'Here, we do things.'

I found Captain's comment irksome.

After returning to shore at the end of my journey, I came across tiny egg cups, made to look like sporting trophies, with the text 'you tried' printed on them. In pointing out that trying is sometimes all you can do, the design, by Adam J. Kurtz, symbolised the exact opposite of Captain's take on life; in which there is no prize for trying.

In the month my fellow crew had spent at sea together before arriving in Tenerife, they started calling the crew quarters before the mast the foxhole, rather than the fo'c's'le. Some called the *Avontuur* a boat, rather than a ship. None of this sat well with Captain, which is why my suggestion that I'd 'try' helming set off a tension that would last the entire voyage.

As I am rather contrarian in nature – annoyingly so, any of my friends will tell you – I could not help but turn this into a joke. I kept using the word 'trying,' and must have told Captain that I would 'convince him' about my use of language. I soon forgot I'd said this. Months later, it transpired that Captain hadn't forgotten at all.

It took me some time to understand our captain. A man in his late fifties with a stiff upper lip, he was somehow both stern and humorous, inflexible and open-minded, blunt and caring. It took me a while to understand how we'd be able to get along.

Meanwhile, I started realising how 'sail cargo' is the antithesis to oil-propelled cargo ships in a dialectical reading of history. In providing a contrast to the enormous ships that ply the seas, sail cargo is not a 'return' to the time before that. It merely offers a contemporary response, inspired by the past. The wind propulsion we see today isn't the same as the wind propulsion of yesteryear. Nor should it be. Wind propulsion follows the same basic principles, as the laws of physics have not changed, but both technology and the global economy have changed beyond recognition. The challenge today is finding ways to incorporate a century of advances in materials, science, and yacht racing technology into a shipping industry that has focused on the optimisation of vessels propelled by burning fossil fuels.

Simply put, if commercial sailing ships had not stopped trading, we would not have had sail training. Sail training as we know it today emerged in response to the disappearance of working sail, in a romantic attempt to preserve its living heritage. The life of 'trainees' cannot, though, be compared to the lives of seafarers before the mast. Without sail training, the appeal of sail would have been far more limited than it is now, leaving little space for sail cargo initiatives to raise funds through paying shipmates.

Traditional 'working' sailing ships ran on the smallest crews possible, to save cost and space. Sail training turned traditional sailing vessels into spaces to accommodate as many people as possible. Cargo no longer paid for professional crews and upkeep; trainees did. Sail cargo offers a blended model, with small paid crews, midsize paying groups of trainees, and small cargo holds.

Despite these obvious changes in economic organisation, life aboard has not changed a great deal. Reading any historical account of sailing large ships, the resemblance with sailing today is striking. Save the *cat o' nine tails*.

We worked in three watches of four hours each. That meant four hours on, eight hours off. With watches changing at four, eight, and twelve o'clock. We were expected at the aft station five minutes before the hour. Twice a day. Every day.

The watch system aboard the *Avontuur* had the added advantage of being conveniently colour coded. The twelve to four watch was green. The four to eight was white. And the eight to twelve was red. Thereby the watches, the navigation lights of the ship, and the clock would overlap with great beauty – though the navigation lights were, as Jonathan Raban observed in *Coasting*, 'cunningly shielded, to avoid blinding the helmsman, and the only visible thing on the *Gosfield Maid* was a weak pinprick of light shining on the compass-heading.'[36]

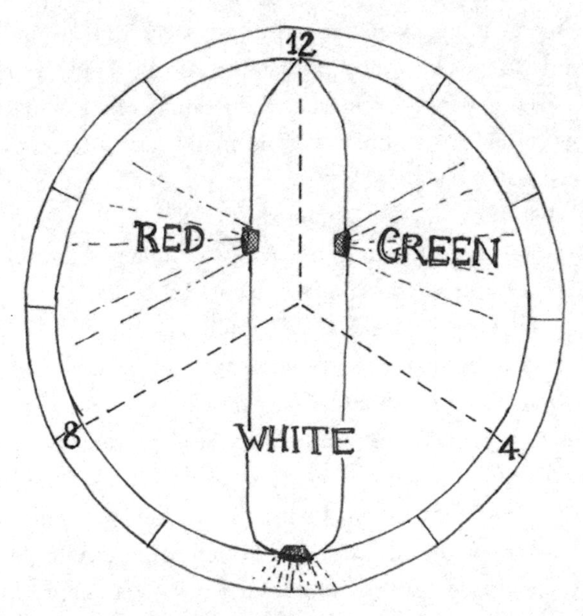

The geometric precision of the overlay of the clock, the ship's navigation lights, and the watch system is striking

Alternative watch systems include two watches of six hours, four hours, or twelve hours each. The one I disliked most was the staggered system aboard the *Tenacious*, on which I sailed from Sydney to Melbourne. Four watches took the deck for periods of two or four hours, shifting throughout the day. While it was a great way to experience the full variety of weather and life aboard, it left me perpetually disorientated, and half jet-lagged. The watch system aboard the *Avontuur* was, in my view, by far the most comfortable one. It left everyone with two blocks of nearly eight hours of free time. One during the day, one at night.

Captain scheduled most manoeuvres so they could happen at watch change. This meant that he waited to gybe or tack, and hoist or reef sails, until two watches were on deck. Only dousing sails – that is, taking them down – often needed quick action, for which the standing watch roped in whoever was around. This made our lives a lot more predictable, as 'all hands' were very rarely called. A manoeuvre delayed the timing of the actual hand-over by as much as an hour, as it normally ends with sending the previous 'watch below.'

One big thing that has changed since the days of commercial sailing vessels is the food. Our Cook was Italian. On a German ship, that's the best you could wish for. Particularly on the first weeks out of Tenerife, we enjoyed fresh food sourced directly from producers on the island. And, for all the struggles we had, Cook cared about food and about us a great deal.

Captain suggested that, in the interest of my research aboard, I rotated watches once a week. That way, I'd have the time to properly meet everyone on board. Social life is commonly, for better and worse, mostly your own watch. After a full rotation of the three watches, I asked Captain if I could remain on his Red Watch, also known as the Captain's watch, as I enjoyed the company of Jennifer and Benji so much – and possibly because the eight to twelve schedule is so pleasant.

My only time alone was in my bunk, roughly the size of a coffin. The distance from my mattress to the top of my bunk was no more than elbow to fingers, with no privacy other than two little curtains that closed off my bunk from the rest of the fo'c's'le. In the tropics, it was far too hot and stuffy to close them.

The curtains did serve an important function. Closed curtains meant 'do not disturb.' Beyond this illusion of privacy, entering anyone's bunk in their absence is strictly forbidden aboard any ship where seafarers' rights are upheld. Aboard the *Avontuur*, it was both a formal rule and a cultural code.

'We like to keep up the fantasy by not attributing anything to anyone in particular,' Bo'sun reminded me. As I got to know him better, I understood that he was asking me on behalf of the entire crew, in his characteristically subtle and understated manner, to keep private things private. We were, after all, aboard

My bunk aboard the *Avontuur*, with view of the fo'c's'le bench, July 17, 2020

a working vessel. Not *Temptation Island* – even so, Timbercoast is affectionately known as Tindercoast across generations of former shipmates.

Bo'sun shared his thoughts on and experiences of other vessels and voyages with me throughout our time together. He was open about ideas and events, but did not speak on anyone's behalf, nor did he blame anyone for conflict or error.

Six professionals could have sailed the vessel shorthanded. But that would have been tough, especially considering the voyage we'd unwittingly embarked upon. Beyond enjoying the pleasures of sail, our presence as shipmates was not accidental.

We were aboard to 'live' the experience of sail cargo and live to tell the story.

We had our personal reasons too. Benji, Mia, and Pinkie left their jobs to travel in South America, which sailing made possible without flying. Jen and Athena were working on a film script about the sea; the voyage offered them a window on that world. Little Captain followed in the footsteps of Alexander von Humboldt, who sailed across the Atlantic by way of Tenerife, intending to climb the Chimborazo and the Teide, in preparation for his book. Peggy joined to learn more about coffee trade flows and the labour conditions of the farmers growing the beans. The voyage would be a major chapter in her journey as a climate activist.

Pol, our Second Officer, had left his job in graphic design years earlier to crew on private yachts. The appeal of community and purpose drove him to sail cargo. Our First Officer joined after a hiatus, as he'd left his work on container ships four years prior.

Captain, too, joined the *Avontuur*'s mission, though somewhat reluctantly so. He was brewing his own plans for a larger vessel, *Quo Vadis*, which would ply oceans as a laboratory for cooperation and shared living.

'Sail is history,' said Captain. 'It's about bringing people together, not transport.' For him, our voyage was more about our encounter and collaboration than about getting cargo shipped from A to B. I'm not sure Cornelius would agree, as the viability of his company relies on getting paid to deliver cargo, not the personal growth of the crew.

Sailing for sustainability

In 1896, the Swedish scientist Svante Arrhenius calculated that halving the amount of carbon in the atmosphere would cool the planet by 4–5°C and doubling the amount would warm the planet by 5–6°C.[37]

At the time, there were about 290 carbon dioxide particles (per million) in the atmosphere, and some 137 million tonnes of goods were transported by sea annually. By the end of the nineteenth century, much of the cargo fleet still operated under sail, as the transition from wind to coal was slow. Moreover, the big upswing of cargo volumes and carbon emissions was yet to come.

In 2022, we're faced with the prospect of a 2.4°C temperature increase by 2100.[38] And that's if everyone sticks to existing policies, if the planned technologies to draw carbon from the atmosphere work. Those are big ifs.

Back when Arrhenius reported his findings, shipping companies did not yet realise their actions would, over the twentieth century, significantly contribute to a warming and destabilised climate. It was not until the 1860s that coal-powered ships became sufficiently energy-efficient to compete with sail. For this to happen, the cost of coal over a voyage had to be lower than the cost of labour to man the sails.

The opening of the Suez Canal on November 17, 1869 meant that coal-powered ships gained a clear advantage over wind-powered ones: the new route not only avoided the perilous Cape of Good Hope, at the southern tip of the African continent, it also halved the distance from Liverpool to Bombay from some 11,150 to 5,750 nautical miles. Most importantly, ships were not allowed to navigate the narrow canal under sail.

In his short history of the Suez Canal, Max E. Fletcher describes how its opening was 'instrumental in giving the steamship an initial entry into one of the last great strongholds of the sailing ship.' But he stresses that 'it required time and considerable improvements in the steamship to complete the job of ousting the sailing ship from the Australian trade.' He concludes that 'it was not until the mid-[eighteen]-nineties,' when Arrhenius published his seminal findings, 'that the sailing ship lost the bulk of the Australian wool trade, and even at the end of the century sailing ships still carried the main burden of the Australian wheat.'[39] It would take two World Wars to advance technology – and shipping capacity – to such an extent that even the legendary windjammer fleet of Gustaf Erikson, on which Eric Newby sailed in 1939, would stop being commercially viable. Erikson was a shipowner based in the port of Mariehamn on the Åland archipelago, located where the Gulf of Bothnia meets the Baltic Sea, halfway between Finland and Sweden. He was one of the last shipowners, if not the last, to use large sailing ships at scale when most had shifted to mechanical propulsion by burning fossil fuels. He kept doing so not for environmental or sentimental reasons, but because he knew he could run a profit by buying the ships on the cheap and sailing them with small crews.

Unfortunately for the planet, the protagonists of empire and commerce embraced the new coal-fired technology. Failing to innovate would, after all, hamper their political and economic dominance. They might be forgiven for not realising the environmental implications of their trade, as their shipping companies unknowingly set course for increasingly rapid environmental destruction. Arrhenius was far from the first to flag the environmental impact of burning fossil fuels. His work built on Jean Fourrier's findings from 1824 that the atmosphere surrounding our planet was what contributed to its stable climate. In 1859, Eunice Newton Foote, an American scientist and pioneer of women's rights, discovered that water and carbon dioxide molecules captured and trapped heat in the atmosphere. (Perhaps unsurprisingly, it was her male contemporary John Tyndall who is widely credited with this discovery, even though he published his work three years later.)

It took politicians a century to take meaningful action in response to Arrhenius' findings. The 1987 publication of Gro Harlem Brundtland's report *Our Common Future* popularised the term 'sustainable development,' which finally turned global warming into a practical problem with a politically feasible solution.

In 1992, shortly after the collapse of the Soviet Union, the United Nations hosted the Earth Summit in Rio de Janeiro, which resulted in the *Rio Declaration on Environment and Development*. It suggested that for the solutions to global warming to be palatable, the future needed to be sustainable for the planet, people, and most importantly, profits. In Rio, UN member states committed to achieving 'sustainable development and a higher quality of life for all people.' They agreed that this goal would require them to 'reduce and eliminate unsustainable patterns of production and consumption,' but not without pursuing 'appropriate demographic policies.'[40]

It took another five years until countries agreed on how to translate these principles into a binding commitment, the Kyoto Protocol, in 1997. The seriousness and scale of climate change was established, but not quite the urgency. The result? Despite lofty ambitions and political pledges, annual carbon emissions have kept on increasing, from 38 gigatons in 1990 to 53 gigatons in 2015.[41] The concentration of CO_2 in the atmosphere reached 420.23 in April 2022.[42] The climate activist Bill McKibben's long-standing target to keep this value below 350 ppm (parts per million) now seems hopelessly naive, no matter how crucial.

Johan Rockström, the climate scientist who defined nine 'planetary boundaries,' suggests we have already crossed several dangerous tipping points.[43] Soon we may be faced with irreversible, run-away warming, as exceeding more tipping points will escalate the climate crisis further.[44]

In the name of economic growth, the shipping industry refuses to immediately adopt an effective short-term option for reducing its role in the problem: slowing ships to their optimal or most efficient speed. Sticking close to such lower speeds, rather than pushing the engine's capacity, is what the industry calls 'slow steaming.' But their refusal to pick even the lowest-hanging fruit means ship operators are lagging in both efforts and results.

Between 1990 and 2008, the carbon emissions from shipping doubled, increasing far more quickly than emissions as a whole.

Since 2008, they have remained relatively stable, but are still far beyond levels the planet can bear.

Take the IMO's 2018 plan to cut emissions in half by 2050 and to zero as soon as possible. Given that the shipping industry's emissions have doubled since 1990,[45] halving them would only mean bringing them back to 1990 levels. Recent Intergovernmental Panel on Climate Change (IPCC) updates make it abundantly clear that this is in no way sufficient.[46]

For comparison, the EU's European Green Deal aims to reduce emissions to 55 percent of 1990 levels by 2030 and to zero by 2050.[47] Meanwhile, the IMO does not seem to grasp the urgency. In any case, it shows little commitment to rapidly increasing its regulatory ambition. Quite the contrary, its 'initial strategy' of 2018 isn't due for revision until 2023.[48] In November 2020, it failed to agree on operational guidelines that would make even the slightest dent in shipping emissions.

None of this is surprising, given that the shipping industry is dominated by a tightly controlled and secretive group of multinational companies, which holds excessive power over its UN regulatory body.[49]

Activists rightly demand radical action in response to the vagaries and double-speak of environmental regulation. More so, these environmental organisations also conduct and publish analyses and research that inform both governments and industry. While Ocean Rebellion and Fridays for Future take to the streets, countless environmental campaigners work behind the scenes to debunk myths, push for greater ambition, and provide data and narratives to help build stronger 'high ambition' coalitions.

Meanwhile, artists try to imagine what a radically different future might look like. In one such near future, depicted in Kim Stanley Robinson's 2020 novel *The Ministry for the Future*, activists sink container ships and shoot planes from the sky to make operations too risky to be economically viable. The wildly fantastic and rather unrealistic narrative of a transport decarbonisation trajectory shows just how difficult it is to imagine how this transition is to occur.

Robinson is not alone in imagining a role for sabotage.[50] And the urgency of the issue has prompted some to call for a rethink of what counts as legitimate climate action.[51] Today, 'Climate activists are sometimes depicted as dangerous radicals,' António Guterres pointed out in response to the release of the latest IPCC repost in April 2022. 'But, the truly dangerous radicals are the countries that are increasing the production of fossil fuels.'[52] Even so, being a radical does not equate to being an eco-terrorist, which hardly ever seems to be effective, except in fictionalised form. There are numerous reasons for this.

Firstly, the unholy alliance of government and corporate surveillance have made illegal acts of sabotage riskier, particularly as the FBI targets environmental activists as 'domestic terrorists.' Will Potter, an American journalist, explains that this has made being 'green' as dangerous as it was being 'red' during Cold War McCarthyism.[53] Few people want to live a fugitive life, like the fictional George Hayduke or his real-life incarnation Paul Watson, who founded Sea Shepherd.[54]

Secondly, the narrative in *The Ministry for the Future* suggests that a few targeted acts of sabotage would be enough to make industry realise that a major shift is needed. Given the long history of piracy, incessant wars, and adverse weather, shipping companies seem undeterred by such risks, as they have not ceased operations. Why would it be different when a few container ships are sunk?

However, the ecological piousness of opting out of plastics, cars, or aeroplanes might feel good, but it is not very effective – not only because you can't consume your way out of climate change. This illustrates that while refusal, or simply opting out, may be a morally superior stance, it rarely works as a strategic option because it's a largely negative act: you opt out. Merely *not* doing things, especially as an individual, will not halt climate change.

Not doing something doesn't make for great policy either. Governments are expected to *do* things, rather than simply avoid inflicting harm. This raises a crucial question: what else are we supposed to do? We must offer a real alternative, showing an alternative that is possible *and* feasible. This requires a wholly different approach. It requires action, not opposition. It requires a *Plan B*, which is crucial to winning a fight against the climate crisis, against the culture that drives it.

Embracing wind propulsion for cargo transport is one such alternative. It can clearly operate without the use of fossil fuels, as it had for thousands of years prior to the discovery of the steam engine. While windmills have made a comeback as turbines generating electricity, sailing cargo ships remain a tiny niche.

When Jorne Langelaan, Andreas Lackner, and Arjen van der Veen looked for investors in 2007, to help them launch the *Tres Hombres* as a sailing cargo vessel, banks and the shipping industry laughed them out of the room. Though in the end, the *Tres Hombres* first crossed the Atlantic for Fairtransport more than a decade before the IMO set an industry-wide emissions reduction target in 2018.

Fifteen years on, sailing vessels, old and new, ply the old trade winds while transporting cargo. Numerous others are being designed, built, and tested. More than ever, the shipping industry symbolises the climate crisis we're in, by reflecting the challenges of decarbonisation, climate justice, and the future of global trade we see in the economy at large within one single industry. Ships, after all, need to decarbonise like every other sector. However, mere decarbonisation does not resolve the thorny question of how to rectify the loss and damage incurred by those who continue to suffer the consequences of colonial exploitation. Nor does it help thinking through the tension between a growth-driven political economy of global trade and a planet that, much like a ship, has clear and finite limits to how many resources can be used and pollution generated without poisoning the well.

The sail cargo movement is now a tiny, though highly visible, player in the struggle to decarbonise the shipping industry. Can these engaged businesses help force the shipping industry to shift back from horrendously polluting, but frighteningly convenient, fossil fuels to zero-emission wind propulsion? And in doing so, will this help mitigate the climate crisis that shows no sign of slowing? Or is sail cargo little more than a symbolic feel-good niche project that caters to environmentally pious consumers who hope to spend their way out of the climate crisis? Meanwhile, how can one explain that the shipping industry has ignored the potential of wind propulsion for so long? Hasn't this technology, which has been around for thousands of years, always been the most obvious way to decarbonise maritime cargo transport?

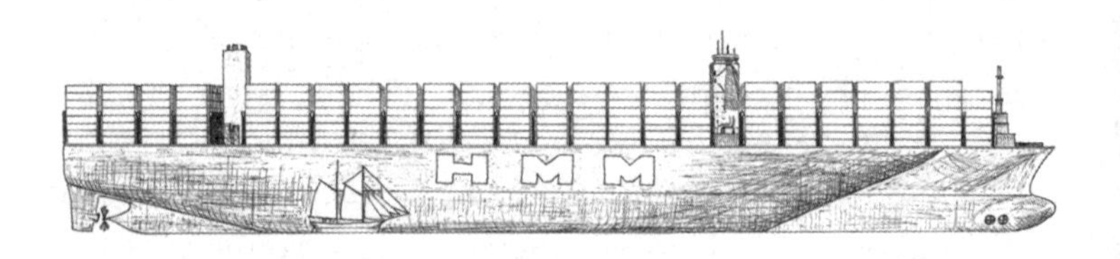

2

What is wrong with the shipping industry?

Shipping is clean, but polluting

Shipping is something I didn't think about much, until recently. That is rather odd, since I grew up in Ghent, a Belgian port city, where I lived on a river barge for half a decade in my late teens. Our living room was in the former cargo hold of the *Nevada*, a thirty-eight-metre-long *péniche*. Half of my family hails from Antwerp, one of Europe's largest ports. My father's family owned a wood import company for several generations. They traded wood from Europe and the tropics, which invariably arrived by ship.

Even so, the penny did not quite drop. I failed to see just how important maritime trade is. I was, as Allan Sekula puts it, 'sea blind.'

I certainly did not consider the aggregate implications of maritime transport on the environment at all. Shipping, I thought, was one of the cleanest ways to transport goods. Per tonne transported, shipping is indeed a good bit less polluting than trucks and trains; its emissions per shipment are negligible compared to air freight. But shipping is a big industry. The emissions of all ships combined exceed those of Germany.

'Shipping is not benign,' Rose George argues, 'because there's so much of it.'[1] And since she published her book, *Ninety Percent of Everything*, nearly ten years ago, shipping volumes have only increased.

Everyone will now be aware that flying is bad for the environment. It is indeed. Though the environmental impact of the shipping industry as a whole is worse. In 2019, before COVID-19 disrupted global travel and trade, aeroplanes contributed some 915 million tonnes of CO_2. Shipping, by comparison, emitted 1060 million tonnes, or just under 3 percent, of global anthropogenic emissions.[2]

In 2014, the Third IMO Greenhouse Gas Study projected that the industry's emissions would, by 2050, grow by 50–250 percent, under various business-as-usual scenarios.[3] In 2020, the Fourth IMO Greenhouse Gas Study revised these projections. The organisation now assumes that emissions in 2050 will come in at between 90 and 130 percent of 2008 levels.[4] This is significantly below the worst projections from 2014. But it remains well above its own emissions reduction target of 50 percent by 2050.

The IPCC warns with 'high confidence' that growth in transport demand 'could outweigh all mitigation measures unless transport emissions can be strongly decoupled from GDP growth.'[5] The Organisation for Economic Co-operation and Development's International Transport Forum (ITF) modelled future demand after the COVID-19 pandemic along three different scenarios. These include 'Recover,' 'Reshape,' and 'Reshape+,' each indicating 'increasingly ambitious post-pandemic policies to decarbonise transport' levels of climate commitments.[6] Despite these 'increasingly ambitious' trajectories, the

ITF predicts significant increases in transport demand, ranging from a more than doubling under a 'Reshape+' scenario to a nearly tripling (from 113 trillion to 312 trillion tonne-kilometres) under a 'Recover' scenario by 2050. While this transport demand extends beyond shipping alone, it offers a good indication that the shipping industry will increase in size considerably, both for the three decades to come and most likely for a long time after that. Given current decarbonisation commitments and efforts, ships are set to become more efficient. Though this 'efficiency is a chimera that leads to the Jevons paradox,' claim the economists Jason Monios and Gordon Wilmsmeier, 'whereby all the efficiency gains in recent times have been eaten up by overall growth, leading to no net reduction of carbon emissions and indeed an increase.'[7]

The IPCC stresses that greenhouse gas emissions from all human activity need to be cut to net-zero by 2050 to have a chance at limiting global warming to 1.5°C.[8] Indeed, the latest available evidence suggests that by 2050, shipping emissions need to be at least 85 percent below 2008 levels, and ideally down to zero – not merely net-zero – if the industry is going to play its part in avoiding excessive global warming of more than 2 degrees.[9] Every sector that can attain zero emissions, as shipping can, should not pursue a 'net-zero' strategy by relying on offsets.

Climate Action Tracker, an organisation that computes the likely temperature increases based on countries' pledges, policies, and actions, considers the shipping industry's 2030 and 2050 levels of commitment 'highly insufficient.'[10] Its calculations reveal that if all countries and sectors adopted shipping's level of commitments to emission reductions, the world would suffer warming of up to 4 degrees by 2050.[11]

In April 2022, the IPCC released its third report as part of the sixth assessment cycle. This report focuses on mitigation, or what we can do to limit the damage. The report suggests that technological options do exist to halve warming by 2030 and cut emissions further by 2050. For shipping, these technologies include e-fuels made from green electricity and innovations such as air lubrication of hulls, more efficient propellers, and of course a variety of wind propulsion technologies. It further acknowledges that 'efficiency improvements [in the shipping industry] can provide some mitigation potential,' but stresses with 'high confidence' that 'additional CO_2 emissions mitigation technologies for aviation and shipping will be required.' In a thinly-veiled critique of the IMO, it flags that 'improvements to national and international governance structures would further enable the decarbonisation of shipping.'[12]

Some companies claim to do better than the shipping industry as a whole. In 2018, the Danish giant Mærsk publicly committed to achieving net-zero emissions by 2050.[13] Two years later, its Lebanese-French competitor, CMA CGM, followed suit.[14]

'The only possible way to achieve the so much needed decarbonisation in our industry is,' according to Mærsk's Chief Operating Officer, Søren Toft, 'by fully transforming to new carbon-neutral fuels and supply chains.'[15]

So far, so good.

Amid the COVID-19 pandemic, the European Commission decided to include maritime emissions in its revised Emissions Trading Scheme (ETS). This did not go down well with Mærsk and CMA CGM. Anastassios Adamopoulos, a journalist at *Lloyd's List*, an industry-focused publication, noted that:

> Mærsk and CMA CGM are looking to secure concessions that would make the ETS rules for shipping less stringent than European lawmakers and environmental groups want them to be, but which would still be more than what many in the industry want, which is for the EU not to impose any emission regulations on shipping at all.[16]

The discrepancy between public commitments to deliver net-zero shipping and the private lobbying for greater allowances and less regulation is telling. While Mærsk has shifted its commitments since, now aiming for net-zero by 2040,[17] the resistance and obstruction by states has slowed down regulation and lowered levels of ambition. I'll later explain how the organisational structure of the IMO enables such freeloader behaviour.

Do shipping companies expect innovations to emerge that will help them to decarbonise without the imposition of exacting regulations that could cost them dearly? Or do they simply hope to be left alone, while setting their own measures at the IMO – which, we have seen, are nowhere nearly ambitious enough?

By building a 'zero-emission' ship propelled by bio-methanol, 'Mærsk clearly wants to present itself as a leader in technological transition for shipping,' comments Faïg Abbasov, the director of shipping at the Brussels-based environmental organisation Transport & Environment. 'But the company is ignoring the basic science. There is simply not enough sustainable biofeedstock to meet the needs of society. Mærsk seems to be ignoring the basic facts about the limits of biofuels because it wants to get a market position by sounding like the leaders.'[18]

Mærsk isn't simply building a new fleet. In June 2021, its CEO Søren Skou called for a fuel levy of US$450 per tonne (equivalent to US$150 per tonne of carbon) to 'bridge the gap between fossil and zero carbon fuels,' but remained vague on its implementation. When *Lloyd's List* requested greater detail, the company said it 'would not be immediate' and 'could start at fifty dollars per tonne of CO_2 in 2027.'[19] Despite the long-term ambition, this makes the proposal less ambitious than the entry-level levy of US$100 per tonne of carbon by 2025 proposed by the Marshall Islands and the Solomon Islands.[20] Meanwhile the International Chamber of Shipping, the industry's lobbying arm, has proposed a carbon levy of a paltry two dollars per tonne of fuel.[21]

Either way, these public commitments do not seem to come with a realistic plan, particularly when considering that Mærsk CEO Skou defended the company's expansion of air freight capacity 'saying it would be customer demand not Maersk's push into air freight that would decide how much cargo was transported by aircraft.'[22] That's right, while Mærsk claims to be a leader in shipping decarbonisation, it invests in the most polluting and most difficult to decarbonise means of cargo transport.

The tension between the abstract target of full decarbonisation and the lack of a clear and realisable pathway also dominated The Economist's World Ocean Summit, held in early March 2021. Shipping industry representatives unanimously agreed on the need for decarbonisation. But none seemed to have a concrete plan to achieve it before the total carbon dioxide budget runs out – which will be well before 2050 at the current rate of emissions.[23]

Why is transforming shipping from a massive polluter into a clean industry so difficult? It is because the shipping industry has dragged its feet for decades. Complacent in its geographic invisibility and economic importance, both the industry and its regulatory body did not take action when it should have in the 1990s.

The Kyoto Protocol tasked the IMO with regulating the industry in 1997, but the UN body failed to set any industry-wide targets until 2018, twenty-one years later. No wonder Greta Thunberg and her fellow youthful climate activists lambast politicians' idle talk: there is so much evidence of it that their empty promises are hard to bear.

One can look at the IMO's 2018 targets with cautious optimism or with cynical pessimism. The three ambitions, central to the IMO's 'initial strategy' are helpful.

The first ambition focuses on bringing down the carbon intensity of new ships. This means incorporating technological innovations and improvements in the design of ships, so they use less energy (and thus fuel and carbon) without compromising on functionality.

The second ambition focuses on reducing the carbon intensity of shipping operations. This means reducing the amount of fuel needed (and carbon generated) per cargo movement, expressed in emissions per tonne-mile (or tonne-kilometre).

Third, this initial strategy also aims to make the aggregate emissions of all ships and cargo operations 'peak and decline ... as soon as possible'. In principle, this means that the IMO aims to 'reduce the total annual GHG [greenhouse gas] emissions by at least fifty percent by 2050 compared to 2008,' with the added promise to pursue 'efforts to phasing them out' such that shipping can attain a 'CO_2 emission reduction consistent with the Paris Agreement temperature goals.'[24]

One could be optimistic, with some justification, by focusing on the commitment-in-principle to align emissions reductions with the Paris Agreement temperature goals – though I'm not quite convinced. My pessimism stems from the fact there are three major challenges with this target. First of all, a 50 percent reduction is not in line with Paris Agreement commitments, which require net-zero emissions by 2050 at the latest.[25] More importantly, however crucial the 2050 target is, the main point of focus should be the remaining carbon budget.[26] Second, the baseline for reductions should be 1990, not 2008, because halving 2008 emissions brings us only back to 1990, which is the basis on which the Kyoto Protocol articulated targets, and on which the European Union calculates its emissions reduction targets under the Fit-for-55 commitment as part of its Green Deal. Third, while the technologies now seem to exist – at least in theory – to fully decarbonise the propulsion of cargo ships, there is hardly any discussion of whether the scale at which we transport goods and consume is 'sustainable' at all. This means that, even if the IMO's ambition were to increase when its 'initial strategy' is due for revision in 2023, along with some countries and companies, we're heading for a future where a fully decarbonised shipping industry continues to grow, in order to support a wholly unsustainable global economy, driven by the consumption of mostly low-quality disposable goods.

Will we sail again?

In 1810, as the Napoleonic Wars raged, King Frederik VI of Denmark realised something had to be done. If his kingdom's fleet was to survive further conflict and adverse weather, it would need plenty of mature hardwood to build and maintain ships.

The problem? In retaliation for Denmark's support of Napoleon, Britain had seized the Danish fleet.[27] At the time, there was a widespread shortage of good timber for shipbuilding, not just in Denmark, but across Europe, making it difficult to rebuild its fleet. In the years between the start of the First Anglo-Dutch War in 1652 and 1862, when 'iron definitely replaced wood as the material for naval architecture,' most countries lacked wood for shipbuilding.[28] 'There was,' until that time, as the cultural historian Aleksandr Ėtkind puts it, 'nothing to replace the floating fortresses built from choice timber.'[29]

Denmark had long had less tree coverage than its richly forested neighbour, Sweden. As early as 1000 CE, only 20 percent of Danish territory was forest, compared to over 70 percent in Sweden. By the early 1800s, when King Frederik decided it was high time to regrow forests, Denmark's tree cover had withered to 10 percent, while Sweden retained a thick cover of almost 60 percent.[30]

Far from being 'sustainable,' the use of timber for shipbuilding has long contributed to the gradual, but steady, depletion of forests around the world. Indeed, at the time it took some four thousand mature oaks to build one single battleship. That many trees require about forty hectares to grow.[31]

Before the 1860s, the worsening scarcity of suitable wood caused navies and the owners of merchant ships great headaches. Today's global supply chains require vast supplies of oil and steel. But before the emergence of such steel ships propelled by fossil fuels, it was the availability of timber for hull planks and masts that could make or break a country's military and commercial success.

'There is a close parallel,' the oceanic historian Robert G. Albion wrote in 1952, 'between the old timber problem and the oil problem of today, except oaks could be grown again in a century whereas oil cannot be replaced.'[32] In saying this, Albion points to a problem that has plagued shipbuilding for centuries. No matter how 'renewable' sources of raw materials may be, they require careful planning and forethought. And culling trees for timber has significant effects on the planet's climate. Indeed, the British economist Nicholas Stern, in his influential *The Economics of Climate Change*, listed deforestation (euphemistically classified as 'land use') as one of the key drivers of anthropogenic climate change. He stressed that 'the loss of natural forests around the world contributed more to global emissions each year than the transport sector.'[33]

When confronted, in 1810, with a badly damaged naval fleet, King Frederik VI decided to have a swathe of oak trees planted in order to avoid Britain's plight caused by a perpetual shortage of boatbuilding timber.

The King designated an area of the northernmost part of *Sjælland* – Danish for Zealand – the country's largest and most populous island as the forest-to-be that would safeguard the future of his navy. The aim was to ensure that, two hundred years on, the Danish navy would be able to build a new fleet from its own mature oak forest.

In 2010, the Royal Forestry Commissioner of Denmark notified Queen Margarethe II that Gribskov, the 'Vulture Forest', was ready for timber harvest, so the Danish navy could build new ships. The thousands of mature oaks that line the forest could serve as excellent timber for this purpose.

The Danish navy, however, has little use for timber in the twenty-first century. The merchant marine has also changed its ways. Mærsk, a Danish shipping company, has one of the largest shipping fleets in the world, but has no use for oak masts.

Few ships today rely on wood for their hulls, masts, or yards. But some do. In the Costa Rican jungle, Danielle Doggett and Lynx Guimond have gathered a team of shipbuilders to construct the *Ceiba* at their Astillero Verde, or Green Shipyard. This wooden ship will become a '100% emission-free' cargo ship for the twenty-first century.[34] Can a ship that needs so many trees to be felled be truly sustainable? After all, didn't shipbuilding strip scores of trees on both sides of the Atlantic? With all the challenges we face today, is building a wooden ship environmentally responsible?

'A tropical tree will capture approximately one ton in twenty-five years,' Danielle told me when we finally managed to speak again in September 2021, well after the pandemic forced me to change my plans to visit the shipyard in 2020.

> If we look at how many tonnes *Ceiba* will be, which is maybe three-hundred-and-fifty, we can say that's approximately … three-hundred-and-fifty tons of carbon sequestered and held within that vessel for up to a hundred more years.[35] And people also look at it in a bit of a more poetic light as well, where they say, 'You've cut this tree, but you've actually given it a longer life if you consider the purpose of the ship as meaningful.' And in addition, the profits of *Ceiba* will continue to plant trees every single year. And that's why it becomes regenerative because the ship, under ideal conditions, will operate easily a hundred years, where these trees that we're cutting are between fifty and seventy years old on average. So not only are we regenerating and replenishing, but we're also increasing the amount that's getting put back into the soil.

We can certainly build one ship on regenerative principles. Or hundreds; perhaps even thousands. But the widespread deforestation of Europe and its colonies shows that even trees, so-called 'renewable resources' because we can cut them down and plant more, have their limits. We can plant more and use wood for construction and shipbuilding, where it can serve as a carbon sink – even though 'sink' is not a word anyone would want to associate with a ship. But no matter how many trees we grow for timber, they won't be able to replace all our needs for concrete and steel to the scale of the twenty-first-century shipping industry.

The *Ceiba* may not be scalable, but it did intrigue me. To find out more, I would travel to Costa Rica, after disembarking the *Avontuur* in the Caribbean, to visit the shipyard to find out how Sail Cargo Inc., the company building the *Ceiba*, deal with this challenge. Due to the COVID-19 pandemic's border closures in 2020, I was not able to visit the Astillero Verde, which Danielle and her team built from scratch on Costa Rica's Pacific coast.

Most of the timber is sourced locally, in Costa Rica, where Danielle told me that the government takes sustainable forestry seriously to the extent that they require two separate permits to obtain a tree for timber; a first one to cut the tree, a second one to transport it. This makes illegal logging a whole lot more difficult. Meanwhile, she continued, 'here in Costa Rica, we've dropped about four hundred and we've planted more than four thousand trees to date, which we do maintain for a period of three to four years to ensure they actually succeed at growing.'

The spars and masts come from Canada, from the Haida Gwaii islands, home to the Haida people for more than 12,500 years. Danielle works with North Pacific Timber, who think far into the future, as their thousand-year forestry management plan illustrates. 'Let that sink in,' Danielle said, 'when people talk about forestry management, or the lifespan of a tree, a hundred years is looking into the future. When we talk about the lifespan of a ship, they're usually insured for twenty-five years. It's planned obsolescence. North Pacific Timber's thousand-year forestry management plan is based on the lifespan of the tree that lives longest on Haida Gwaii, the Kiidk'yaas or Golden Spruce, which is approximately a thousand years.'

Growing trees for shipping requires foresight. King Frederik VI of Denmark learned that the hard way. That's why I ended up speaking to Sam Faucherre and his team who are restoring the wooden ship *Hawila* in the Danish port of Holbæk.

The *Hawila* is a 1935 gaff-rigged galeas, which served many purposes throughout her long life that began in Sweden. Since 2014, she has occasionally served as a sailing cargo vessel, combining the transport of goods in the Baltic Sea with 'an onboard educational platform to inform about the issues surrounding globalisation, and challenge the current food system culture.'[36] Like the *Avontuur*, she sails with a purpose: showing that another, greener, kind of cargo transport is possible.

The ancient Greeks knew all too well that wooden ships required a great deal of maintenance. The realisation that ship maintenance requires the replacement of its rotten timber led to Plutarch's riddle: if Theseus' ship has – over its lifetime – every plank, beam, and mast replaced, is it still the same vessel, or is it a wholly different one?[37]

The *Hawila*, much like Theseus' ship, eventually requires all of her timber to be replaced as rot can be delayed, but never fully stopped. If the Hawila Project has an explicit environmental mission, then how can its new keelson and spars be sourced sustainably?

When the ship was due for a major refit in September 2020, she was lifted onto the dry in the town of Holbæk, deep in the Isefjord. Thanks to the foresight of King Frederik VI, the committed work of the *Hawila* team, and *Naturstyrelsen*, the Danish government's Nature Agency, she was able to use fifteen of the now two-hundred-year-old oak trees planted during the King's reign, to replace her rotten planks, keelson, ribs, and deck beams. The *Kagerup Savværk*, a sawmill near Gribskov in operation since 1889, cut the timber to size after the Hawila Project negotiated the use of the oak trees.

'It makes me happy,' says Peter Chrois Møller, a forest ranger for *Naturstyrelsen*, 'that at least some of our great ship oaks can fulfil their destination [*sic*] and become part of a great sailing ship.'[38]

The number and size of merchant vessels has increased to such an extent that the required timber harvest for shipbuilding would hardly be sustainable. It would increase the rate of deforestation to even higher levels than are seen today. Shipping has increased more than five hundredfold since the early 1800s. Given the levels of deforestation caused by shipbuilding at that time, the planet would simply lack the space to grow all the trees needed to build and maintain the number of ships in use today.

Unfortunately, the alternative has caused other problems. The shift to steel coal-powered ships did not eliminate the challenge of securing resources; it merely created a need for various ores and fuels. The growth of international trade and the wide availability of cheap oil has turned the shipping industry into an essential part of global supply chains. While today's ships use no trees for their hulls or masts, their excessive use of fossil fuels causes great harm.

Many environmentalists call for 'rewilding' by having trees, plants, and animals take over the planet's surface again.[39] Planting trees is key to drawing down carbon and ensuring habitats for wildlife. Planting trees and protecting wildlife is crucial because humanity's survival depends on it. Perhaps counter-intuitively, cutting down trees from well-managed forests and turning them into wooden ships can make those ships into carbon sinks that also help to decarbonise the shipping industry. Clearly, wooden ships aren't scalable to meet all the needs of the twenty-first-century shipping industry. This is partly because timber can't be used to build enormous ships, but also because shipbuilding wouldn't be the only industry vying for sustainable timber.

Sea blind

Allan Sekula, a photographer who spent his career documenting all things maritime, jotted down a simple but telling calculation in one of the many notebooks he filled during his decades-long work on the shipping industry: One 'Super Panamax' ship, with capacity for seven thousand twenty-foot containers,[40] burns so much Heavy Fuel Oil (HFO) that it pollutes as much as all the road vehicles of London.[41]

Sekula is, with good reason, not the only one worried about the environmental impact of such ships. This is the pollution caused by just one – now smallish – ship. There are more than sixty thousand vessels in the global merchant marine. That adds up to fifty thousand Londons worth of air pollution, just to ship our consumer needs from mine to factory, from factory to store, and from well to bowser.

A sensationalist *Daily Mail* headline, *How 16 Ships Create as Much Pollution as all the Cars in the World*, combined a frightening statistic with a very vivid image.[42] Fred Pearce, a regular contributor to *New Scientist*, refers to research conducted by Jim Corbett into a particular kind of pollution, sulphur oxide (SOx), which is emitted when burning fuel with high sulphur content.[43] Large cargo and cruise ships have long emitted far greater amounts of SOx than cars because fuel regulations for cars are tighter than for ships, which mostly use HFO. More than three-quarters of the fuel used by ships is HFO, a toxic sludge so dirty that no one else will buy this by-product of oil refineries.[44]

Regulating the use of HFO is one of the success stories of the International Maritime Organization. Sort of. Since January 1, 2020, ships have been required to burn fuel with low (less than 0.50 percent) sulphur content (down from less than 3.50 percent), or install 'scrubbers,' which eliminate SOx particles from exhaust fumes.[45] This 'sulphur cap' cuts particulate matter that harms health, particularly near major shipping lanes, but also further afield as these particles can travel great distances.

There are, however, some major downsides when using scrubbers.

The 'sulphur cap' increases carbon emissions, as the scrubbers used to 'clean' exhaust fumes use a lot of energy – and thus more dirty fuel.[46] These scrubbers 'wash' the fine particles that are no longer permitted from exhaust fumes. But what happens to the wastewater this creates? Most ships simply flush it into the sea, thereby creating spikes in ocean acidity. This is terrible for marine life as acidity dissolves sea creatures' calcium carbonate skeletons and shells.

If the scale of pollution from shipping seems baffling, that's because it is.

In 1840, around the time when Richard Henry Dana Jr sailed cow hides from California around Cape Horn to Boston, the world's seaborne trade weighed a combined twenty million tonnes.[47] This added up to less than twenty kilos of goods per person per year at that point in time. By 2018, we were shipping more than eleven billion tonnes, five hundred and fifty times more.[48] The world's population increased less than tenfold in the same period.

The total amount we now ship per person? Nearly one and a half tonnes. That's a lot of trainers.

The historian Frank Trentmann explains in *The Empire of Things*, his excellent chronicle of consumer society, that humans are now primarily *consumers* of goods and services. This explains why many people think that their individual consumer choices can save the planet or, at the very least, halt climate change. Trentmann explains in meticulous detail how what we choose to consume is important; consumers have a role to play. But he also stresses that individuals cannot do much about the 'choice architecture' that influences how we consume – that is, 'their own inertia, procrastination, or unfounded optimism' that influences how we citizens behave as consumers.[49] Consumer capitalism promises empowerment through consumption. But consumer capitalism won't let itself be consumed, at least not in the way that things are consumed by fire. That is why he concludes that 'our lifestyles, and their social and environmental consequences, should be subject to serious public debate and policy, not left as a matter simply of individual taste and purchasing power.'[50] While lifestyle changes are necessary, we can't simply rely on individual consumer choice to accomplish this.[51]

Despite the growing reliance of economies – and people – on maritime transport, shipping has largely disappeared from public view. Ports have moved from inner city waterways and shorelines to gated compounds on city fringes. We've become increasingly 'sea blind' to the scale of the shipping industry. Out of sight, out of mind. Like most people today, I long failed to grasp the scale of the global shipping industry and the enormous carbon emissions it produces.

Public attitudes to these emissions have changed in recent years. Research about air pollution near major ports shows that air quality is impacted by ports and shipping lanes.[52] Sensationalist headlines like the one above, along with those lamenting the abominable air quality aboard cruise ships, have caused alarm.[53]

Improving air quality by regulating the particulate matter emitted by ships is simple. Agreeing on a global sulphur cap was easy, compared to regulating carbon emissions. The far more difficult task faced by the shipping industry is full decarbonisation.

Since the industrial revolution, shipping has gradually shifted from sail to coal; and later from coal to oil. As economies globalised, as international trade increased, and as offshore oil extraction became the norm rather than the exception, cargo ships increased in size. They now ship almost everything we consume at least once, often multiple times – first as ores, fuels, and grains, then as consumer goods, and ultimately as waste.

Tack to the future

'Do we really need to just improve what we've already got?' Cornelius asked when we first met in Elsfleth. It was mid-October 2019 and I had just arrived in the small town on the western bank of the Weser, halfway between Bremen and Bremerhaven, after a long train ride from Berlin, via Hannover and Hude. Elsfleth is one of these German towns that feels utterly removed from the rest of the world, as if stuck in an era that never existed anywhere else.

'We've got a very efficient shipping industry with modern ships,' Cornelius told me that evening over dinner, answering his own question, 'and we can make these ships clean by different kinds of propulsion, like sails, but do we really want to do this, or should we rather take a few steps back and reduce everything?'

His question is not limited to shipping. We stand at a major fork in the road. Do we double down on innovation to maintain an ideological commitment to economic growth? Or do we take a step back and reduce consumption, resource use, and transport, by gently 'degrowing' the economy? Do we invest and invent more, to chase the promise of modernity? Or do we downsize, reduce the size of the economy by choosing to live simpler and slower lives?

Cornelius' position, turned into practice through the *Avontuur*, is simple: 'Go back to natural materials. Go back to a lot less transport. Of course, a lot less consumption. A post-growth economy.'

But how does this deep-green, anti-consumerist perspective relate to global attempts to reign in the carbon emissions from the shipping industry through regulation? And how does the promise of *degrowth* chime with a company that aims to sell its goods and grow its 'ethical' and 'green' consumption as a way out of the current climate crisis?

Despite their obvious differences, sail cargo companies resemble the American outdoor clothing manufacturer Patagonia. This upscale brand targets environmentally aware consumers to buy fewer but better products. They charge a premium for producing goods that are ethical (not made in sweatshops), green (made from recycled materials and thus 'good for the planet'), and durable (by creating quality gear and by offering repair and resale options to limit consumption and waste).

I have certainly fallen for their marketing. I love the idea that what I buy will last. This is, of course, to a great extent a self-fulfilling prophecy. Buying something pricey with the intention of using it for a long time might mean you care more for it, making it last longer.

Ironically, these anti-consumerist ideas are good for business. By marketing their products as a way to consume *less*, Patagonia actually sells *more*. Ethical consumption has won popular appeal and market share.[54]

Some time ago, Patagonia adopted 'Don't Buy This Jacket' as their Black Friday slogan.[55] But in pursuing this anti-consumerist marketing strategy, they sell an image. Anyone wearing Patagonia, myself included, engages in virtue signalling. 'I bought this lightweight waterproof jacket because I *need* it,' my appearance screams when strolling in drizzle, because I 'do not waste my time or money on pointless things.'[56]

'It is a double greenness,' Sharon Hepburn argues in her analysis of Patagonia's ethos – or is it marketing? – 'it is not only the greenness of production processes to protect nature, but also greenness as a more ethereal, aesthetic quality with parallels in Western literary discourses, that links the wearer with images of the sublime *in* nature, and the sublime experience *of* nature.'[57] Beyond the greenness of the production of the brand's clothing, Patagonia sells membership of a group of environmentally conscious adventurers.

Like many young retirees escaping the incessant toil of business, Cornelius could not simply sip drinks by the pool in sunny Queensland all day. He wanted to do something. Live a little. And make a difference.

The plan was simple. He would run a sailing cargo vessel along the Queensland coast, combining shipboard family life with a shipping business that would carry both goods and tourists on coastal routes.

He told me this story over dinner at the Kogge, one of the few restaurants in Elsfleth. With some ten thousand souls, a maritime academy, and a wharf, the town became the *Avontuur*'s home port. He never intended to move back to his native Germany. And even less so to the Weser estuary where he grew up. But after a long search for the ideal vessel for his shipping plans, he found the *Avontuur*. That search wasn't easy. Once acquired, she needed a great deal of work before she'd be able to carry cargo across the oceans. New masts and rigging were crucial. But so was reclaiming space for cargo, while allowing enough space for sixteen crew.

The *Avontuur* would continue her long history as an ocean-going cargo ship. After all, she served as a sailing cargo vessel for most of her life. When Paul Wahlen sold the vessel in 2004, its new owners transformed it into a charter vessel for day sails on the Dutch Waddenzee. Hence the need for a serious refit.

After obtaining quotes from several shipyards, Cornelius decided to work with the Elsfleth shipyard on the Weser. Once completed in 2016, the shipyard's bill had ballooned, and Cornelius was broke. Rather than immediately returning to Australia, he decided to sail an Atlantic round trip to replenish his funds. Today, returning to Australia is no longer part of the plan. Having completed many Atlantic round trips for Timbercoast, the *Avontuur* is now at home in northern climes – as is her captain, Cornelius.

In 1867, the British adventurer John MacGregor set sail on his *Voyage Alone in the Yawl Rob Roy*. Jonathan Raban, who later coasted around Britain in his *Gosford Maid*, reflected that by the time of MacGregor's adventure 'there was no domestic wilderness left.' By the mid-nineteenth century, British adventurers were deprived of land-based sublimity and turned to the seas for transformative encounters with the sublime. At sea, it was possible to live away from industrialised civilisation and the pollution it caused.

Allan Sekula argues that the sea of 'exploit and adventure' has now turned into a 'lake of invisible drudgery.'[58] It's therefore hardly surprising that most people simply don't care about seaborne trade. Books about maritime history celebrate the exploits and adventures of people at sea while books on contemporary shipping try to bring an otherwise invisible world to a specialist readership.

Sail cargo companies let their consumers experience the maritime sublime. Through their social media channels, sail cargo companies convey the 'exploit and adventure' their crews live through in order to deliver coffee and rum to ethical consumers in Europe. Or better, they invite those consumers to join their vessels on the next voyage to experience the hard and hazardous work required of sailors from yesteryear as they delivered luxuries from the erstwhile colonies. Few take up such opportunities. But many celebrate the commitment of sail cargo crews as they brave the stormy North Atlantic, the wintery North Sea, or the muggy tropics. The careful marketing of goods 'shipped by sail' share the strategies perfected by Patagonia. First, sail cargo companies differentiate by claiming qualitative superiority of the rum, coffee, and chocolate they sell. Then they insist that their way of working is inherently green, by using the cleanest means of production or transportation. After that, they implore their consumers to consume less; by stressing that if they *must* consume, they should opt to consume *their better* products. Finally, they sell the experience of adventure, which many are keen to buy into, but few are willing to join in person. They prefer *stories* about gale-force winds through which the rum they sip by the fireplace has travelled.

Navigating the perpetual tension between having to act as an individual consumer (for want of real alternatives in daily life) and making strategic demands about shifting society as a whole (because individuals won't be able to consume their way out of this mess), leaves us with tricky questions of how to ensure our acts have impact. In this context, young climate activists rightly claim that politicians and business leaders do little more than kick the can down the road. Ambitious promises and commitments, like the Paris Agreement, mean little if they are not met with equally ambitious actions.

Kim Stanley Robinson's book, *The Ministry for the Future*, mentioned earlier, defends the future of unborn generations. The titular Zurich-based intergovernmental agency, founded as a result of the Paris Agreement, engages with politicians, UN agencies, and business leaders. Its director, or 'Minister,' Mary Murphy, embodies the values that Mary Robinson, Christiana Figueres, and Laurence Tubiana have been fighting for.[59] Murphy hopes to ensure that the myopia which characterises politics does not impede meaningful long-term climate action.

The Ministry accomplishes its goals through a strategic blend of legal action, diplomatic back-channelling, tactical politicking, and covert 'black wing' action. Convincing national banks to back a Carbon Coin as a means to fund climate action through quantitative easing is very effective.

Unfortunately, *The Ministry for the Future* is a fiction. And future generations remain unheard in climate debates – though change may be on the horizon. In September 2021, António Guterres published *Our Common Agenda*, in which he proposes a UN Special Envoy for Future Generations.[60] This, Thomas Hale argues, could help us deal with the 'long problems' we face today.[61]

My *Avontuur*

Joining the *Avontuur* above any other sailing cargo vessel came down, like pretty much everything in life, to serendipity. When I first saw Cornelius at the station in Elsfleth, I knew he'd understand my interest in sail cargo. Several meals and conversations later, he understood my research angle and was keen to have me on board for the upcoming voyage.

He was to sail the ship on her fifth Atlantic round trip since he bought her. I planned to join for just a few weeks of the six-month round trip, from Tenerife to Marie Galante. When we made those plans in October 2019, I committed to three weeks aboard. I simply couldn't take more time off work, I told him. But in this short time, I hoped to catch a glimpse of what sail cargo was all about.

But the *Avontuur* was not my only option for an initial experience of sailing cargo. The *Tres Hombres* had already been plying the Atlantic route for several years. The *Kwai* was shipping cargo under sail between Pacific Islands. Meanwhile, the *Grayhound*, a replica lugger launched in 2012, shipped cargoes mostly across *La Manche* or the Channel, as the English call it. Then there was the *Apollonia* on the Hudson River or the *Providence* that plies the Strait of Georgia out of Vancouver. Perhaps The Blue Schooner

Company's *De Gallant* or the *Hawila* in Denmark would do. Options galore.

I'd be lying if I said that the ship did not matter. Compared to the *Tres Hombres*, the *Avontuur* is sheer luxury. Never mind that the latter lacks showers and heating too. The ship is brighter, roomier, and on the whole more comfortable. On top of that, the two companies exude very different cultures. Jennifer and her daughter Athena looked into both the *Tres Hombres* and the *Avontuur* before deciding which one to sail on. Much like me, they thought the former a bit more hippie, crusty, and feral; and the latter more neat, outdoorsy, and rigid.

Whatever the reasons, the *Avontuur* is the ship I ended up sailing on. Her story is now integral to my own. Not that any of this will stop me from boarding the *Tres Hombres* one day, if they'll have me.

Entrepreneurial types like Jorne, Arjen, and Andreas (the *Tres Hombres*), Cornelius, and Danielle thrive on dreaming up big visions. They have to, or no one would go along with their out-of-this-world plans. Without committed crews, their dreams would flounder on the docks.

When I first set eyes on the *Avontuur*, she looked worse for wear. A rainy October morning in northern Germany did little for my first impression. She was as cold and damp as the German autumn. She felt empty and lifeless.

But that's the reality of ships: they need crews and the open sea to give them life. It's only when living aboard that you realise just how alive and animate a ship can be.

This two-masted gaff-rigged schooner was built by Otto Smit's shipyard in Stadskanaal near Groningen in the Netherlands. In operation since 1920, the ship's rich history includes use as a sailing cargo vessel between 1980 and 2004, under command of the late Captain Paul Wahlen, who died in 2008.

When Cornelius bought the vessel ten years later, she'd been transformed into a charter vessel sailing the shallow Dutch

Waddenzee. The rum and coffee on board were there for immediate enjoyment.

The *Avontuur* is forty-three and a half metres long, if you include the bowsprit. On deck, she was roughly the same size as the *Nevada*, a five by thirty-eight river barge, on which I lived in Ghent for five years. The *Avontuur*'s hold is considerably smaller than such a *péniche*, though. Even so, her living quarters are designed to accommodate sixteen people, rather than a single household.

She typically carries a sail area of 495 square metres, rising to as much as 612 when the topsails fly. That is around the size of two tennis courts. These sails propel her 114 deadweight tonnage capacity. Her 135 cubic metres of cargo space is ample but, depending on the stowage factor – or weight to volume ratio – the hold cannot quite fit all the weight she's allowed to carry.

On the aft deck, exposed to the elements, is the helm. Directly in front and some seventy centimetres above the deck, stands an island of instruments and hatches. A bit higher still, is an instrument box housing a compass, engine controls, a rudder-angle indicator, and a GPS-driven speed indicator. The latter failed before we finished our Atlantic crossing, leaving us dependent on chart room speed readings from the ECDIS chart plotter.[62]

Left of the instrument box is a hatch leading into the 'aft peak,' where Bo'sun kept his non-edible treasures, mostly manila rope and pine tar, giving the space the exquisite aroma of marlin twine. Throughout the voyage, Benji and I tried to emulate the smell, by creating a Bo'sun n°5 scent, for which Athena designed a fitting label. What else do you give a Boatswain for his birthday? Just forward of the hatch leading into the Bo'sun's domain is one of two entry points to the engine room.

Below sits a 224 kW John Deere main engine, which can be fired up for emergencies and manoeuvres. The engine room also houses the water maker, a fire pump, and – the bane of my existence – a big and noisy generator that tops up the batteries

The Electronic Chart Display and Information System (ECDIS) in the chart room, March 15, 2020

stored in the engine room. The twenty photovoltaic panels on the aft cabin's roof did not supply enough power to meet all our electricity needs, including essentials like the navigation equipment and water maker. The *Silentwind* windmills mounted on the stern taffrail were, however, neither silent nor functional.

Along a companionway and down a few steps, you entered the deckhouse. On port, you found the chart room. On starboard the aft heads. Forward was the passageway to the crew's cabins – where Captain, the officers, Bo'sun, and Cook lived; with a modest library suspended above the staircase. Just fore of the heads was the passage to the galley and mess. As the *Avontuur* has no separate messroom, we referred to the single space as the galley. This lack of separation made the galley the beating heart of the ship. Courtesy of the tempest named Giulia – our Cook – that space was often as tense as it was cheerful. Our only certainty upon entering the space was that we'd immediately know what kind of mood our Cook was in.

'The boat,' the late French philosopher Michel Foucault says, 'is a floating piece of space, which exists by itself, that is closed in on itself and at the same time is given over to the infinity of the sea.'[63] Every ship contains a parallel social universe that does not quite exist before or after you embark. The culture on board

View from the 'helm box,' at anchor off Puerto Cortés, April 11, 2020

is a uniquely temporary constellation that will only exist for a single voyage. In his reflection on 'heterotopias,' or 'other spaces,' Foucault explores how suspending many of the rules of life ashore can give rise to experimental social relations aboard a ship. At sea, this liminal suspension of ordinary social relations is both necessary and easy.

This is the paradox of the 'open' ocean. The water surrounding the ship appears boundless, but in fact functions like an oversized moat, enclosing the seafarers aboard. The ocean feels open and inviting, as long as you do not *want* to leave the ship.

A ship forces a wholly different form of social life onto its crew. Within no time at all, a group of people who've never previously met is required to form a close community, built on discipline and skill, with a common purpose: moving the ship and staying alive.

Without the money to pay crews, sail cargo pioneers need volunteers to restore, maintain, and sail their ships. Rather than signing on for need of work and payment, we joined for a purpose. The professional crew got paid. But they could have made more money with less effort. The shipmates like me paid in time and money to be there.

We relived at once the lives of seafarers of yesteryear and the uniquely privileged perspectives of Eric Newby and Richard Henry Dana Jr. What separated us was a long time and different worlds we left behind. But we joined the same benevolent dictatorship, in which we submitted, *nolens volens*, to Captain's will. Our safety, as the anthropologist John Mack reminds us, 'depends critically upon [our] submission to a common purpose.'[64]

That purpose is, at the very minimum, the ship's integrity. The vessel needs to stay afloat so it can transport crew and cargo to their destination. The captain's responsibility is to ensure everyone works to this end. Once, when I was sailing around Argyll and Bute in Scotland, that ship's skipper dryly remarked that the ship always comes first in an emergency. If you can't save the ship, there's no point saving the crew. While the captain will do everything in their power to ensure the safety of the crew, they will take fewer risks than anyone else. Our common purpose extended beyond merely staying afloat; we were there to bring Timbercoast's 'mission zero' to life. Through collaboration, we'd explore if and how we might live and work together for a future of the shipping industry, the global economy, and – rather modestly – the planet as a whole.

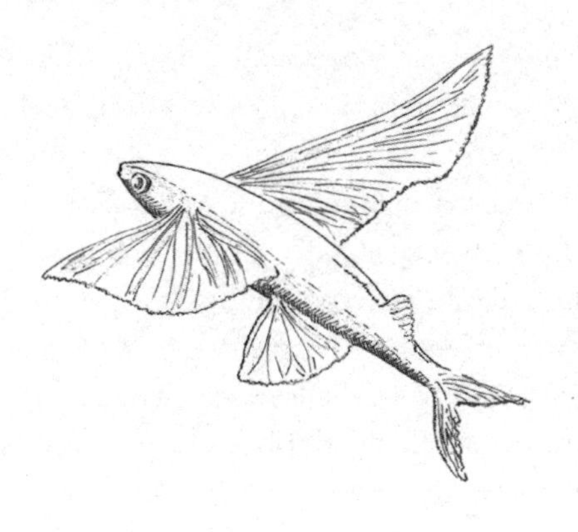

3
Crossing the Atlantic

Life at sea is hard, but unfair

I am not one for manual labour. No one aboard the *Avontuur* would disagree. Physically idling is in my nature. This habit is made worse – better as I see it – because the intellectual labour that is my bread and butter, remains largely unseen.

No such excuses count aboard a sailing vessel. All hands are needed to keep the vessel going and shipshape. I enjoy the skill of sailing and the craft of ship maintenance and repair – though I can't get behind the culture of looking busy, which I hoped to escape. There I was with my preference for an Arendtian *vita contemplativa*, wholly at odds with the *vita activa* that is sailing.[1]

As a kid, I rejoiced in finding the book *How to Be Idle*, by Tom Hodgkinson. My mother gave it to me as a present. Not that I needed the encouragement.

This contrasts with age-old sailing culture, which celebrates busyness and activity at all times. Not solely for practical, but also for political purposes.

'It has always been a maxim with me to engage and occupy my men,' said Lord Admiral Collingwood, 'and to take such care of them that they should have nothing to think of for themselves beyond the current business of the day.'[2]

We had little time or space for ourselves on *Avontuur*. The legacy of centuries of sailing, during which common seafarers were kept obedient by keeping them busy, continued to influence life at sea. It was, captains thought, the idleness of their 'men and boys' before the mast that fomented unrest and mutiny. Hard work served not only to keep the vessel shipshape, but also to keep workers under the thumb. Controlling labour on ships remains crucial.

'Life at sea is hard,' quipped Bo'sun, 'but unfair.'

Benji soon started repeating that sentiment more often than anyone else aboard, much to Captain's annoyance. Captain firmly believed that one quickly starts believing what is often repeated. He would, for example, insist we refrain from using 'no problem' to mean 'you're welcome.' The mere mention of the word 'problem,' he thought, would make us think of problems, which might end up causing them. He had a habit of eradicating bad habits at their root, which in his view seemed to be language. 'We can do it together' should replace 'only I can do it,' Captain told us, a bit more didactically than was strictly necessary. He equally insisted that 'I am useful but interchangeable' should trump 'I am irreplaceable.'

Benji was only half joking when repeating Bo'sun's joke over and over. He first heard about the *Avontuur* when watching a German TV documentary on a lazy Sunday afternoon at his Shanghai apartment, where he was working for a German company. He was immediately keen to join and organised to do so at the end of his eight-year stint in China.

He expected a more leisurely and comfortable environment. Not quite a cruise, but certainly much less work and more time for social and private life. Benji was by no means lazy. Quite the contrary. He was one of the most diligent shipmates aboard. Even when his relationship with Cook soured over time, as her fits of passion clashed with his cheerful stoicism, he kept on cleaning the galley with great care every night after she turned in.

As an engineer who grew up on a Bavarian hobby farm, he fixed many things I considered well beyond salvation. A sextant, for example, requires an accurate timepiece, which is why Captain bought a stopwatch in Tenerife to replace the ship's one, whose battery had died. Finding a new battery on the island proved more difficult than buying a toy stopwatch costing fifty euros. Then again, the grossly overpriced stopwatch we picked up in Tenerife stopped working within weeks. Benji and I took to the task of fixing it.

'Heiland,' Benji cursed, when we dropped a tiny screw the first time.

'When you say Heiland, you mean Jesus, no?' I asked. It means the same in my native Dutch, but wanted to make sure I understood. Every German on board knew exactly what Benji meant, but no one else ever used the expression.

With the sérieux of the Muppets' grumpy duo Statler and Waldorf, we spent an entire afternoon trying to fix the stopwatch that screamed planned obsolescence. We lost the same tiny screw not once or twice, but seven times before Peggy told us she had spare batteries that would fit the good stopwatch. We gave up and replaced the battery instead.

Peggy packed spares of everything, because she did not quite know what to expect. She'd taken all this equipment for her *Peggy Merkur* podcast. It didn't help that she was unaware that there would be proper bunks aboard. Neither did she realise how the watch system would work.

Not that Timbercoast or Cornelius were to blame. All voyage brochures explained the organisation of both time and space at sea.

It is, however, probably easier to understand these things when you have previously sailed on a 'sail training' vessel. I remember how clueless I was before my first tall ship voyage on the *Tenacious*, despite their detailed joining instructions.

'Are not sailors very idle at sea?' asks Richard Henry Dana Jr. No, he responds, 'the discipline of the ship requires every man to be at work upon *something* when he is on deck, except at night and on Sundays.' Indeed, he continues, 'it is the officers' duty to keep everyone at work, even if there is nothing to be done but to scrape the rust from the chain cables.'[3]

Officers do far more than keep people busy. They also take on the responsibility which they have in the ship's chain of command. My academic titles, on which I rely ashore, are worthless at sea. I signed on as a sailor before the mast, and that would be my role. I had to follow orders. Not something I am used to doing.

One of the deckhands liked this idea even less. He thought himself captain, for that's what he was aboard his own vessel. Never mind that he signed on as a deckhand; he wanted to be in charge. That's why we started calling him Little Captain.

Our Cook had previously sailed on the *Tres Hombres* and never failed to remind us that the *Avontuur* was a luxury cruise by comparison. Not only because the *Avontuur* has more space, more fresh water, and more electricity aboard, but also because it is far easier to sail a schooner than a brigantine, as the former has fewer sails and does not require climbing aloft to set or douse the latter's square sails.

Meanwhile, some of my shipmates needed no encouragement; they had internalised the idea that toil equals virtue. Joni, a most diligent worker, seemed to relish deck work and general maintenance. I'd prefer to look at the ocean with a cup of tea.

Being at the helm, unlike rust-busting, gave me a sense of immediate usefulness. To me, fighting rust is necessary, though the rhythmic hammering on steel is nowhere nearly as satisfying as steering a large sailing ship. It was also the only place on deck where lowly shipmates like me could exercise a modicum of control over our environment. An unwritten rule dictated that whoever was helming could select music, played through a wireless Bluetooth speaker on the aft deck. The reasoning, I understood, was rather sensible. The person at the helm was the only one who could not leave their station, so they should not be forced to listen to someone else's musical taste.

Fossil fuels conveniently diminished capitalists' reliance on labour.[4] Though, rather than freeing up labour from coercion, it led to more capital gains for investors and more consumption. I'd hope that shifting to renewable energy will free up people's

The deck of the *Avontuur*, March 12, 2020, Atlantic Ocean

time and finally make good on the promise of technology to give us freedom. What, indeed, happened to the fifteen-hour working week John Maynard Keynes promised us in 1930?[5]

Nothing of the sort has been accomplished today. The more productive we have collectively become, the more we work. I can't fathom why. I believe we should embrace technology to save time – though not to be more productive, but to be idle. That, to me, is part of climate justice. Some indeed argue that being less productive, or at least not chasing productivity gains, is key to living within planetary boundaries.[6]

I loathe the glorification of labour. I find myself far more aligned with the French-Cuban Marxist author Paul Lafargue, who wrote *Le Droit à la Paresse* (*The Right to be Lazy*) in 1880. 'The machine is the saviour of humanity,' he argues, 'the god who shall redeem man from the sordid artes and from working for hire, the god who shall give him leisure and liberty.'[7]

Sail cargo, and many other small-is-beautiful craft movements, do the opposite. They reclaim time from mindless labour through time-consuming labour. This is a common trope in degrowth thinking, where labour, once decoupled from economic coercion and the need for work to survive, would turn into a pleasant and entirely voluntary activity.[8] But let's not pretend that reclaiming time from the rat race is easy. Even if it were easy to replace bullshit jobs, as David Graeber calls them, is fun *work* really the solution?[9] Especially when bearing in mind that small-scale sail cargo can't really replace all our maritime transport needs, at what point would automation take over and turn the craft of sailing ships into observing automation, which mightn't be fun work either?

I'm not claiming that the craft of sailing has no point at all. I greatly enjoy the patience and skill involved in making gaskets from bits of rope, whipping loose ends, and repairing this or that. Having a patient Bo'sun explain these crafts helped, especially since he did so with a creative and problem-solving mindset, rather than a dogmatic rule-based one. He never pretended to know the best way to do something. He rather encouraged us to explore solutions, while giving us pointers when doing so and being open to learning from any new pathways or methods we could come up with. As the voyage progressed, it was precisely our working together that taught me a thing or two about how we can tackle the climate crisis. I'll come back to that later.

My personal enjoyment aside, I could not stop wondering: why don't we have fully automated sailing ships yet? Small crews could operate vessels that harness both the wind and the technological innovations that have occurred since the heyday of sail in the nineteenth century. A range of newly built sailing cargo ships aim to incorporate these labour-saving innovations. Some major shipping companies aim to build new sailing vessels for cargo transport, but they remain in various stages of 'technological readiness.' So, despite a history of thousands of years of sail, commercial shipping has failed to seriously consider this option for decades. Zephyr & Borée's *Canopée*, under construction at the Nepture Marine Shipyard in Szczecin, Poland, TOWT's two eighty-metre 1100-tonne vessels under construction in France, Veer Voyage's Dyna-rigged ship, and the Oceanbird project by Wallenius Wilhelmsen and the *Neoline* are most advanced in their development.

These newer 'primary wind propulsion' initiatives mostly aim to operate far more within the existing shipping industry, rather than create a parallel market, by providing a tiny niche driven by shipping tiny amounts of luxury goods by sail.[10]

Except the *Canopée* and TOWT's ships, which should be in the water by the time this book hits the shelves, these 'primary wind' vessels could take years before being deployed commercially. The majority of 'zero-emission' vessels and propulsion technologies pursued by the industry will rely on 'green' hydrogen or ammonia to fuel ships. In theory, this can work, but no such technology has been tested at scale. The portside infrastructure to facilitate such a shift is also lacking. In other words, while the environmental impacts of burning fossil fuels have been known for more than a century, the shipping industry has made precious little effort to instigate a shift to 'green' propulsion technologies. Given the average twenty-five year lifespan of commercial cargo vessels, the entire current fleet is all but 'locked' into another few decades of polluting fuels, leading to high 'committed emissions.'[11] This is why retrofitting the existing fleet, ideally with wind-assisted propulsion technologies, is so crucial; it would save emissions now and thereby buy time to get to zero as an industry, as it would extend the remaining carbon budget.[12] All the while, the necessary ships, as well as shore-based facilities, to safely and reliably replace toxic oil burned by the tonne, simply do not yet exist.

Even so, at least one such ship could have been in the water by now, if the project had not been abandoned in favour of labour-intensive traditional sail. Shortly after the *Tres Hombres* set sail in 2007, Fairtransport worked with the Dutch naval architects at Dykstra to design the *EcoLiner*, a cargo vessel that would use Wilhelm Prölß's 1960s Dana Rig sail design. The Amsterdam-based firm Dykstra was first put to use for the *Maltese Falcon*, an eighty-eight-metre-long yacht, in 1990.

Meanwhile, despite extensive refits, the traditional 'sail cargo' vessels that are in the water now do not even use electronic winches. If we had used such electric winches, Jennifer would probably never have burnt both her hands on the main sheet during a gybe in early April. Even so, with them, we would not have gained muscle or skill.

Machines, I'll gladly admit, can be noisy. Every time the diesel-powered generator turned off, I felt a sense of relief. The ensuing silence was blissful, even if the noise was not annoying while it was there. It was only after it stopped that it became obvious just how loud and disturbing it was. It was as if the sky immediately became calmer. For its lack of luxury, the *Tres Hombres* had the greatest appeal of all: silence.

The 'tres hombres,' the three men behind Fairtransport, complemented each other in terms of skills, which allowed them to refit a 1943 ship for commercial cargo transport, tell convincing stories, and build a profitable company. But their differences led to tensions, as they never quite agreed on a unified vision for an alternative future. Jorne's insistence on building a traditionally rigged *EcoClipper* over a newly designed and fully automated *EcoLiner* drove tensions to breaking point. While all three of them remain shareholders, and thus co-owners of the vessel, only Andreas is involved in the everyday running of the vessel. Jorne focuses on the *EcoClipper*, while Arjen hopes to revive the *EcoLiner*. The tension between the futures they imagine boils down to what they think about scale and the balance between labour and automation. There is little room for nostalgia and craft on the *EcoLiner*, where very small crews could operate the vessel at the press of a button, without leaving the bridge.

Danielle Doggett of Sail Cargo, the company behind the Costa Rican ship *Ceiba*, has dug up, radically revised, and rebranded the *EcoLiner* design by Dykstra. Under a new company, Veer Voyage, she presented their plans to build six vessels with a capacity of one hundred containers (TEU), significantly fewer than the 476 containers that would have fitted on the *EcoLiner*. She presented the plans and design at the Zero Emission Ship Technology Association event in the fringes of COP26 in Glasgow.[13] Her intention is nothing short of beating all major shipping companies by having a zero-emission ship, propelled by sails and green hydrogen, in the water before anyone else does.

While the *Canopée* should be in the water first, this vessel won't attain zero emissions, as it will use a fossil fuel powered main engine. In Nantes, Neoline is raising funds for a car carrier, while Wallenius has started working with Alfa Laval, in Tumba near Stockholm, to perfect the technological specifications of its *Oceanbird* vessel. Meanwhile, TransOceanic Wind Transport in Le

Havre plans to build a fleet of thousand-tonne automated sailing freight ships.

The race is on.

Perhaps if Fairtransport had stuck together and pursued the *EcoLiner*, the modern and fully automated 8000-tonne or 476-TEU ship might have been at sea by now.

What characterises these new initiatives, in contrast with traditional sailing vessels, is their pragmatism. A commercially viable scale for a zero-emission vessel trumps a sense of community and craft aboard. But what, on such a vessel, would remain of the social experience of working together that is so valuable? This was precisely Jorne Langelaan's concern.

'We are in fact creating a new world by doing business differently,' Langelaan told me about the *EcoClipper*. He stressed that a vision for a different kind of society is needed: 'We do not focus on adapting to what exists, we are constructing a wholly new way forward.'

Even so, in early 2022, Langalaan announced EcoClipper's purchase of the 1912-built ship *De Tukker*. This flat-bottomed sideboard coastal ship will serve the North Sea, carrying some fifty to seventy tonnes of cargo and up to twelve paying guests.[14] Raising the funds to build a new, but traditionally rigged, ship have proven difficult so far – though raising funds for fully automated sailing ships isn't necessarily easier.

Idle talk

'Avast,' Captain yelled while a gybe was going awry. It was March 9, well over a week into our Atlantic crossing.

'Avast,' he shouted again. No one heeded his call. Not for wilful refusal or malice. We simply had no clue what he meant. In his commitment to upholding the culture and language of sail, Captain's jargon was at times lost on us. It was as if we were working towards the same ends but could not agree on the necessary language to accomplish them. This tension reminded me of the mismatch between the careful and cautious language of the Intergovernmental Panel on Climate Change (IPCC), particularly in its first reports, and the right-or-wrong soundbites politicians thrive on. No matter the urgency, the language through which we describe what we do and how we do it makes or breaks our actions.

'Winning an argument,' which politicians seem to care about, 'became a substitute for discovering the truth,' reflects Theodore Zeldin in a little book on the importance of good conversation. 'Forcing others to agree became the source of self-esteem. Rhetoric became a weapon of war.'[15]

The scientists whose work feeds into the IPCC's reports, as well as the scientists who compile and draft these lengthy documents, are not trying to score political points. They are not spin doctors trying to engineer a political win or horse-trade a deal. Their nuanced messages keep falling on deaf ears.

Aboard the *Avontuur*, we focused on getting things done, by responding to commands. After every manoeuvre, we met on the aft deck to ensure everyone was safe. We then also discussed what we'd just done and what could go better next time.

Despite everyone's best efforts, the fragmented work of pulling sheet or halyard, sweating and tailing, belaying and coiling left many aboard wondering about the bigger picture. It wasn't until months later when we started taking a step back, by looking at manoeuvres, rather than doing them, that we started seeing how our own actions could improve.

We understood the jargon, picked up nuance, and grasped the implications of each minor action. We started to better understand the totality of the action, rather than just remembering a small part of the job at hand.

When it comes to climate change, few step back from their station to see where Ship Earth is going. Some may hear Captain calling 'avast,' though few seem to heed or even fully understand the call.

'In the text of the Paris Agreement,' Amitav Ghosh argues, 'there is not the slightest acknowledgement that something has gone wrong with our dominant paradigms; it contains no clause or article that could be interpreted as a critique of the practices that are known to have created the situation that the Agreement seeks to address. The current paradigm of perpetual growth is enshrined in the core of the text.'[16]

'Avast,' in case you're wondering, means 'stop.'

By the time Ocean Rebellion took to the streets in August 2020, the IMO had started translating their aim to decarbonise the shipping industry into practical proposals. Because of how the UN body works, as explored in greater detail later on, member states propose certain pathways for regulation, which they then seek support for. In this case, Japan proposed to extend an IMO regulation for newly built ships to existing vessels. Their proposal would extend the Energy Efficiency Design Index, in place since 2013, to the Energy Efficiency Existing Ship Index from 2023 onwards. It may sound helpful to apply environmental regulations for new vessels to those already in use. This would, you might think, help reduce emissions.

The problem? It probably won't.

The problem is that the regulation focuses on marginal increases in carbon intensity, not total emissions. This is precisely why Ocean Rebellion and other environmental action groups, such as Pacific Environment and the Clean Shipping Coalition, which includes several NGOs such as Transport & Environment, Seas at Risk, and Ocean Conservancy, have raised concerns. While the shipping industry claims to be the least polluting means of transport, a senior Environmental Defense Fund campaigner, Panos Spiliotis, told me he could scarcely believe how inefficient the shipping sector is compared to aviation. Despite shipping industry claims of being more efficient and cleaner than any other mode of cargo transport, there remain simple steps that can help to immediately bring down emissions.

They are not alone, as an independent report on the implications of the proposed measures, prepared by the International Council on Clean Transportation, also concludes that 'the package of short-term technical and operational measures which includes reducing vessels' engine power and the introduction of ship-level carbon intensity targets will only shave off 1% of the sector's emissions growth by 2030.'[17]

In other words, hidden behind the lofty commitments of IMO member states is the ugly truth: these regulations are in no way sufficient to reduce the shipping industry's emissions to zero. As the graph in figure 10 illustrates, the projected reductions in 'carbon intensity' won't help to bring down total emissions at all. Worse, they'll barely help to curb the growth of total emissions, which are likely to increase further.

Despite this evidence, industry representatives such as Guy Platten, the Secretary-General of the International Chamber of Shipping, claims 'the shipping industry is firmly on track to meet the ambitious IMO CO_2 reduction targets and ultimately be a zero-emissions sector.'[18]

This level of public relations spin conceals three major weaknesses with IMO plans.

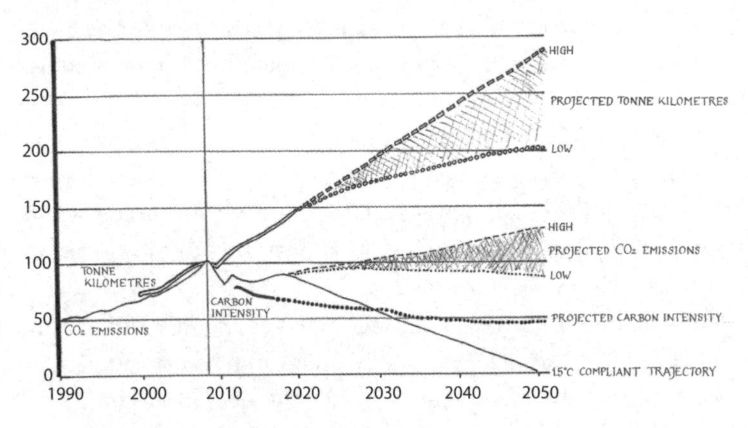

The future of the shipping industry? (based on IMO 2020 data)

First, the IMO's target simply isn't ambitious enough to limit global warming to 1.5°C. This means that it will fail to meet the commitment set out in the Paris Agreement.

Second, the IMO lacks a realistic action plan that will force the shipping industry to meet its plans to halve emissions. A trajectory to fully decarbonise the industry is even less realistic.

Third, the 'operational measures' IMO member states have thus far managed to agree on will not make a dent in total emissions. In fact, it is far more likely they will allow emissions to grow for at least another decade.

Considering the workings of the organisation, it could seem like a great win that the IMO agreed on a 50 percent reduction target. Coming up with such a weak target so late in the game and based on a 2008 baseline, no less, is a significant step – even though it does not align with the Paris Agreement temperature target. Now, both the IMO and the shipping industry can publicly proclaim to be taking action. While its member states happily pay lip service to this target in public, most are reassured that whatever they push for behind closed doors won't get out.

The world as you know it no longer exists

On March 19, everything changed. After nearly three weeks at sea, cut off from news, phone, and email, Captain received a message via satellite.

'The world as you know it no longer exists,' said Captain upon reading Cornelius' message to us on the aft deck. What no one thought possible had happened. Practically the entire world had gone into a hard and strict lockdown to curb the spread of SARS-CoV-2, a novel coronavirus.

When the virus ripped through Wuhan months earlier, the city went into a strict lockdown. We knew that before setting sail. But 'experts have questioned whether the quarantine measures are helpful,' read *The Guardian* at the end of January in 2020.[19]

Australia quickly closed its borders to travellers from China, though they were initially still able to travel via third countries as long as they spent fourteen days there. It remained hard to imagine that Wuhan-style lockdowns would soon sweep the globe.

Just before we left Tenerife, one case had been detected in the Costa Adeje Palace Hotel, on February 25. This plunged the thousand guests into sudden quarantine in their rooms. Tourism in the resort town on the south of the island ground to a halt. Though none of this impacted the carnival that was in full swing in Santa Cruz.

We knew something was brewing, though I never expected all borders to slam shut while we were at sea. And yet, there we were, marooned aboard the *Avontuur*.[20]

All ports and anchorages in the Antilles were closed. Pretty much all of Europe was in lockdown. The situation was deeply uncertain on all fronts. It was unclear whether any of us would be allowed ashore – though that question was largely moot. When we received Cornelius' update, it was doubtful whether any shipmates would be able to travel to Guadeloupe to allow crew change. Without replacement crew, no one could disembark either. As we signed on to sail the ship, not to simply get a passage, we'd have to stay on to sail the ship and uphold Timbercoast's mission, rather than leave the ship stranded half a world away from Elsfleth. That is what it felt like to me, at least. The integrity of the crew was essential to the functioning of the ship and Cornelius' project.

As we inched closer to the Caribbean, it remained unsettled whether those who planned to disembark would be allowed to leave the ship. And if so, what were our onward travel options, if any?

I could not help but think how odd the situation was. As a citizen of the European Union, I joined a German-flagged vessel on the Canary Islands, off the Moroccan coast. We were sailing across the Atlantic to the Caribbean. But legally speaking, we had sailed from Spain to France. Even while at sea, we legally remained in the European Union. And yet, we were refused entry into Guadeloupe, a French overseas territory, even though everyone aboard had the right to live in the EU. Born after the 1985 Schengen Agreement, I'd always considered the whole of the European Union home.

Since our passage exceeded the incubation period of the coronavirus – and the quarantine duration imposed to combat the spread of COVID-19 – it seemed odd that we were being turned away.

Clearly, the world as we knew it really had indeed ceased to exist. The next day, we received an update from Cornelius. We would probably call at Pointe-à-Pitre, but certainly not in Marie Galante. This meant the ship would load provisions to sail onwards, but not the planned cargo of rum barrels.

At that stage, there was no clarity whatsoever on whether we'd be allowed to disembark. Captain assured the three of us – Jen, Athena, and me – who were scheduled to leave the vessel with him in Marie Galante, that he would under no circumstances leave us in a worse situation than we had aboard.

That left only two possibilities. Either we'd travel back to Germany where the voyage was scheduled to end, or we'd stay on the *Avontuur* until the coast was clear, whichever came first.

I was hopelessly solipsistic. I thought about me and my plans. I was meant to travel to Costa Rica and onwards to South Africa. None of this was going to work out. I'd have to notify the University of Melbourne, my employer, that it might be a while longer before I could return to teaching.

While I empathised with my own shipmates, I didn't quite realise that the sudden closures of nearly all the world's national borders meant that one and a half million seafarers were left stuck on cargo ships, not even counting the many more sailors and fishermen who were also out at sea. We were but fifteen of them. Both history and the 'crew change crisis' show how precarious and dangerous maritime labour is. Life at sea, as Benji once again reminded us, is hard, but unfair.

'Nature will take over again,' Peggy dreamily commented. 'With human activity slowing down, nature will restore herself.' She hoped that lockdowns would serve as a push to change behaviour and help tackle climate change. To save nature from the incessant onslaught of the economy, the environmentalist's credo goes, we need to restore nature's balance. This means more nature and less economy. More sailing, less trade. Half the planet for 'nature' the other half for humans, as Edward O. Wilson suggests.[21] But is it really that simple?

In *The Parallax of Growth*, Ole Bjerg argues not.[22] He begins by pointing out that ecology and economy are both thought to tend towards an equilibrium. But if that's the case, then why is life on our planet so 'out of balance'?

The novelist Amitav Ghosh observes that since the nineteenth century, nature has increasingly been seen as 'moderate and orderly' – much like economists see the market. It may fluctuate, but it is intrinsically orderly.[23] This notion, rooted in modernity, is now challenged by climate change.

In a span of less than three weeks, decades of globalisation seemed to have come undone. Rather than promoting open borders – mainly for goods and capital and rich people – the world had slammed shut. The late sociologist Ulrich Beck's warnings about the *World at Risk* had materialised once again.[24]

In the early morning of March 24, when we reached Pointe-à-Pitre, we had to face the New World. All hands were called on deck at 05:30. We dropped anchor. Within hours, the harbourmaster called us on to come alongside.

Weighing the anchor makes for good exercise. The manual winch needs two people on either side, turning the flywheels with screw-on metal handles. While Captain stayed aft with someone at the helm, everyone else was on the foredeck taking turns to bring in a hundred and fifty kilos of steel – without even counting the cable chain. Bo'sun kept a close eye on the winch mechanism, which required constant tweaking to avoid the movement slamming its brakes shut. The Second Officer closely observed the chain coming in, looking out for the anchor. Over the radio he relayed progress to Captain, who kept the engine on stand-by. All in all, it was a quick job, because the remaining eleven of us managed to seamlessly rotate the winch.

We had no idea what to expect when navigating into the tropical port under the grey and humid morning sky. This was not the Caribbean arrival I'd hoped for.

'Just as Schrödinger's cat is both dead and alive until the box is opened, the world ashore is both in crisis and normal until we make landfall,' we wrote in a 'crew statement' just before arriving. 'Given the limited information we have now, we have only our imagination to build an understanding of life beyond our ship. We are not sure when or where we will be able to leave the ship to find out how the world around us may have changed.'[25]

From a safe distance, wearing a surgical mask and latex gloves, the harbourmaster asked us to stay aboard and await the delivery of our provisions. Gerard Petrelluzzi, the ship's agent, remembered Captain from earlier port calls on different ships.

A small truck showed up to deliver the food. We washed it in a solution of water and vinegar. We had no idea how infectious this new virus was, though we were intent on keeping it ashore.

By 11:45, the harbourmaster, who'd long since left, called Captain and asked us to depart. By noon, we'd cast off our mooring lines and were heading to La Ceiba in Honduras.

As we sailed, it was difficult to comprehend that, beyond our isolated community of fifteen, the world was spiralling into a full-blown pandemic.

'It felt so calm and peaceful on deck last night,' I wrote on March 28. 'Barely any clouds. A sky full of stars. And a crew silently gazing at it all.'

Leaving Pointe-à-Pitre after a few hours in port, March 24, 2020

Even so, we had no respite from our time at sea or the unending watch system. By March 31, we had received confirmation that cargo operations would happen in Honduras, Belize, and Mexico. There would certainly be no crew change in the first two ports, though Veracruz seemed a remote possibility.

Throughout the pandemic, the economy remained functional. Barring the occasional pre-lockdown run on toilet paper, food remained on supermarket shelves, tools and medication arrived as ordered, and raw materials and fuels made it to their destinations. It was largely thanks to the seafarers, whose lives aboard extended from port to port, that the world remained the same, even if arriving back ashore proved to be a shock for those who spent much of 2020 working at sea.[26]

I quickly realised that the pandemic would become a costly challenge for Timbercoast. Delays meant that cargo fees would not be paid until much later. They are, after all, due on delivery. The lack of shore leave for restocking food supplies also generated significant additional costs as ship chandlers charge exorbitant fees on top of inflated food prices.[27]

Most importantly, Timbercoast would lose revenue from shipmates who could not join due to travel restrictions. As the company's stores of *Avontuur*-branded coffee, rum, and chocolate from earlier voyages started running low in Elsfleth, cash flow was starting to dry up. The longer this voyage lasted, the more difficult it would be to keep the company afloat. Despite this added pressure, Cornelius stressed that things were safer for us aboard than for people ashore. We'd have to watch out when loading cargo, as contact with agents, stevedores, and port officials posed a risk to us. There was no risk from others on board, as we were virus free.

Even so, the mood started to shift after leaving Pointe-à-Pitre. Our voyage was no longer voluntary. We were confined to sea, as those ashore were to their houses. We wondered how long it would last. On March 27, I jotted my 'best bet' down in my notebook: 'not before we get to Faial, Azores (current ETA June 10).' I'd be both right and wrong.

Infinity

Every Sunday, at 16:00 sharp, all hands gathered on the aft deck for a special occasion: the Captain's Reception. It was usually the only moment during the entire week when all fifteen of us came together. More importantly, for some, it was the sole moment in the week during which the consumption of alcohol was allowed. It was regarded as an oddity that I turned down the opportunity.

The *Avontuur* is a dry ship – 'German flag, no fun,' Benji would repeat more often than funny – so a sip of rum made for a welcome change to the ship's never-ending routine.

This recurrent social gathering suspended the ship's hierarchy. We were all on friendly terms, but we knew who was in control. For one hour every week, we'd uphold a pretence of equality by ignoring, ever so briefly, that Captain was in charge and we had nothing to say. It was the reconnection to the pleasures of shore-based life (a drink, mostly) and snapping out of the routine and order of shipboard life that allowed us to 'reset' our social relations and gear up for another week at sea.

Cook always made 'plum duff' in the form of a more contemporary and luxurious treat than any nineteenth-century sailor ever saw on board. One Sunday, surrounded by Sargassum – the sea-grass that floats through the sea – she wanted to serve us *Prinzen Rolle* (*Prince Biscuits*). To her surprise, there weren't any left in the dry store.

The Red Watch didn't understand the fuss, nor the mystery. We knew all too well where the cookies had gone. Many an evening, Benji would dip into the ship's stores to liven up our evening watch.

Cook didn't take the disappearance of the food lightly.

We, the Red Watch, were in the wrong. Food stores are the cook's responsibility. Dipping into the limited supplies without their knowledge is not done on any ship. Cooking three times a day for fifteen people is difficult enough. But planning meals for fifteen people with limited supplies makes for a difficult life when you can't restock. Knowing what's in store is essential to keeping everyone fed.

Cook was dead certain I was the culprit. I am, after all, a 'lazy and greedy motherfucker.' Her words. But this time, rather than hurling insults, she simply gave me the shoulder.

As much as pilfering food is unacceptable, Cook seemed to be concerned about more than this. We bore the brunt of her frustration with the pandemic and with it the uncertainty of how much longer we'd be stuck aboard; how much longer until she'd see her partner again.

I could not help but think of the ensuing drama of simmering conflict about some snacks as a metaphor. An ethical dilemma played out in miniature. How do we act when faced with limited resources? Do we follow rules? Do we keep stores a secret? Do we divide things equally? Or is everything up for grabs? Encouraging sensible and responsible consumption is part of the *Avontuur*'s raison d'être. Signing up in principle is easy. But what does scarcity do to our own convictions and habits? And should we have had these decidedly non-organic, sugar-heavy, highly processed snacks on board in the first place? Should we have opted for organic-only supplies? Should we have bought only fair-trade chocolate ashore? Should our personal snack choices have been healthier? The question of food was existential.

We had started worrying about food supplies, for it wasn't clear when and where we'd be able to replenish our stores. We received conflicting messages about which ports would or wouldn't be open for business. And there was no clarity whatsoever about the accuracy of the information we had. It was 2020, after all. Everything could change overnight.

So, we were told to ration food and gas. Be careful. Be spartan. Do more with less.

Later that day, Cook made a dish in the gas-guzzling industrial oven.

'Rationing gas?' Benji and I quietly quipped. 'Let's make an oven dish!'

There was little actual dialogue aboard about what we really needed and what we could easily miss. But that's shipboard life. There is a clear chain of command, and little democracy. We had little say, but no responsibility either. Which is precisely why stealing food is such a taboo.

Between standing watch twice four hours a day and eating, we had to fit in sleep when we could. While part of the Red Watch, I was in my bunk from roughly midnight to 07:00, leaving me with some six and a half hours of sleep. Nowhere near enough for me. So, I napped after lunch for another two or three hours. And sometimes I had a quick lie down just after dinner.

My daytime naps coincided with the Cook's 'day shift.' No wonder she thought of me as incorrigibly lazy. Despite my innate laziness, that wasn't the reason I slept so much. The lack of privacy and shade meant that my bunk was the only place where I could come to my senses and come to myself.

One such afternoon, on April 1, I was napping after lunch. Joni came to my bunk and woke me up in his characteristically gentle bedside manner.

'Come aft, the pool is open.'

When I showed up on the aft deck a minute later, donning my swimmers and goggles, I could not quite believe that our incessant joking about building an 'infinity pool' on deck had paid off.

The result far exceeded anything Benji and I had dreamt up. We thought of building a tub on deck with a tarp for walls. Bo'sun had a better idea. The Caribbean Sea would be our pool.

Now, this may seem odd to anyone who's never been sailing. But one does not simply go for a swim when a ship is at sea. Michael Phelp's two-hundred-metre freestyle world record (just under one minute and forty-three seconds, in case you were wondering), translates to a speed of just under 3.8 knots. Needless to say, I am no Olympic medallist. Anything faster than a knot or two would mean the ship would sail on, while I'd become shark fodder.

Bo'sun craftily attached a cargo net to the stern of the ship. The pilot ladder dropped off directly from the stern taffrail. This allowed us to 'swim' while sailing. Or rather, we were able to get our skin exfoliated by the manila rope that kept us attached

The infinity pool, seen from the aft deck, April 2, 2020, Caribbean Sea

to the ship. Bo'sun, who convinced Captain it was safe to use the contraption, vicariously enjoyed the refreshing 'swims' in our mobile pool. Even so, descending into the infinity pool was not without risk.

'Any remark passed on the deck of a vessel seventy-eight feet three inches long,' thought Captain Jack Aubrey in *Master and Commander*, 'was in the nature of a public statement.' So, too, our every whisper aboard the *Avontuur*. The architecture of the vessel made it impossible to know for certain that no one would be able to hear you. So, you'd always be ready to break off a conversation mid-sentence and continue it hours, or days, later.

The view from the infinity pool, April 2, 2020, Caribbean Sea

While in our infinity pool, Benji, Jennifer, and I noted that Captain was doing well, given the circumstances. We did, however, wonder how well Little Captain was coping with the ever-lengthening voyage.

Given the noise of the water around the hull of the ship, we thought we'd found that unique spot where our conversation was private. We were wrong.

Captain, Bo'sun, and Joni heard every single word from deck.

No one else called the deckhand Little Captain. It was our private observation about the dynamic of the erstwhile Red Watch, which was no longer private. It's indeed hard to find time for yourself aboard a ship. And it's even more difficult to find the space to develop one-on-one conversations. No matter where you are, anyone could join your conversation without warning.

After more than a month at sea without a break, a gybe went badly one morning. Bo'sun was visibly unhappy that we still didn't know the drill. It felt as if that urgency of getting things right still hadn't sunk in with the entire crew. I couldn't help thinking: if we can't manoeuvre smoothly in calm seas, what will we do when the weather gets rough?

'At sea,' Eric Newby remarked dryly, 'one was very likely to find oneself let down by one's own mistakes.'[28]

By March 31, it was clear that all planned cargo operations would go ahead. Never mind border closures, cargo transport cannot stop. But there wouldn't be any shore leave or crew change in Honduras and Belize. Some of us cherished hope that crew change would be possible in Veracruz, Mexico. But most had accepted that 2020 would be a year unlike any other in living memory. Perhaps naively, I hoped that we'd ride out the pandemic at sea.

4

Coffee, rum, and chocolate

First cargo in Puerto Cortés

Close to midnight on April 7, we dropped anchor off Puerto Cortés, Honduras. The smaller port of La Ceiba, where we had planned to call, was closed due to the pandemic, so we headed to the country's largest container port.

The smell of land, that typical scent of burnt tropics drifted from the green hills onto our anchorage. As we lay anchored in the protected bay, the lack of movement was unsettling at first. It's hard to get used to one's living space going from perpetual movement to very slight rocking in a sheltered bay.

We took turns 'showering' with a hose on deck under the cover of darkness, using fresh water from a ballast tank that had been in the hold for the Outward and Middle Passages. It was my first proper wash since leaving Gran Canaria, well over five weeks earlier.

There was a palpable uncertainty about whether and where crew change, repatriation, or shore leave would be possible. At some level, we cherished hope that things ashore would not be all that bad. Even so, hardly anyone thought it a real possibility to simply leave the ship and head home.

After all, the ship needed crew. Without the necessary hands, she would be unable to continue. So, our being permitted to *leave* the ship was only half the challenge. Replacement crew in identical numbers would also need to be allowed to *join* the ship – after travelling to Puerto Cortés in the first place. Cornelius asked us not to leave the ship, even if we were allowed by port authorities, as long as no replacement crew was present.

Captain reiterated his promise not to leave anyone in a worse place. Our ship was a haven from COVID-19 as long as we limited contact with shore. The real difficulty was not so much that we would not be granted shore leave or crew change, but the possible arrival of new crew members who were far more likely to carry the virus of which we knew precious little in early April 2020.

We spent most of Wednesday April 8 waiting for port authorities and medical personnel to clear us in. In the meantime, our yellow 'Q' flag signalling our request for *pratique* to enter port, remained up.[1] Not until all checks were completed and paperwork filed would we be given 'free pratique,' or the permission to enter port.

The next day, we lay at anchor, waiting for permission to 'come alongside' and load our first cargo. Captain recounted how the legendary Felix Nikolaus Alexander Georg Graf von Luckner, Count Luckner for short, had inspired him and his father to build their own adventures in life. The Dresden-born captain of the *Seeadler* (*Sea Eagle*), who served the Imperial German Navy during the First World War, earned the nickname *Der Seeteufel* (*the Sea Devil*) for himself and *Die Piraten des Kaisers* (*the Emperor's Pirates*) for his crew.

Cornelius seemed surprised at how well we were simply accepting the situation. He jokingly called our time on board a 'psychological experiment.'

'I am the researcher,' I quipped, 'documenting the experiment, no?'

'Well, haven't the others found out yet?'

'They did make jokes about my being a researcher at first. But I turned out to be as big a weirdo as they were. I managed to throw them off the scent.'

Using Captain's *Skyroam* internet connection, we were able to take turns connecting our phones and pulling in our messages and emails. Sail cargo ships have a strong online presence, for they mostly sell an idea rather than goods. Beyond the chatter and carefully crafted images, the time and energy spent on making cargo transport under sail happen can scarcely be captured on Instagram. Our lives and work were mostly offline, very direct and involved. There was just generally less talk and more action.

The next day, once all the paperwork was sorted, Holger, our ship's agent, called. Cargo operations would start by noon. There was a sense of excitement, verging on disbelief, that we were about to do what we crossed the ocean to do: load cargo. This is precisely why people like Bo'sun are here. Talk is one thing, but what matters is action. So, let's get some cargo.

We prepared ship. We had to move Bob, the tender (a rubber boat used to connect the *Avontuur* to shore, when anchored), from the top of the cargo hold into the water, then take the thick burgundy tarp off the cargo hold to lift the wooden planks that close off the hold. Then we lifted the heavy metal beams, which support the wooden planks, by hoisting them up using the working lines attached to the masts. After storing the last spare ropes and tools in the aft peak, the hold was ready for stevedores to line with Kraft paper.

From that point onwards, all we could do was sit back and wait for the stevedores to load and stow the cargo, while counting the bags to check their number against the cargo manifest.

First cargo, lifted on board pallet by pallet, each carrying fifteen 68-kilo bags, April 10, 2020, Puerto Cortés, Honduras

As the bags of coffee trickled into the hold, the *K-WAVE*, a British-flagged container ship, berthed right beside us. With a capacity of seven hundred twenty-foot containers, she's small by almost any standard. But even next to such a dwarf, the *Avontuur* felt incomparably small to me. Being so much bigger than us, she carried vastly more. In half the time we spent in port, she came in, offloaded cargo, and left. Cranes lifted dozens of containers, each carrying more than we spent an entire afternoon loading, as if they were matchboxes.

Our being a 'general cargo' ship made cargo operations difficult, even with the help of a crane. Few ships need stevedores for manual loading and unloading the way we did. Malcolm Purcell McLean invented the shipping container in 1956 to make the slow work of stevedores easier – and the process cheaper for shipowners.[2] Dockworkers begrudgingly accepted this change, realising full well that they could not stop technology.[3] Containers soon transformed the shipping industry, enabling it to offer lower prices, which in turn made offshoring the manufacturing of consumer goods possible.

The *Avontuur* shipped sixty five tonnes of goods in 2020. Peanuts compared to the *HMM Algeciras*, which the Korean shipping company Hyundai Merchant Marine (hence HMM) launched in April 2020. At that time, it was the largest container ship in the world, with capacity to carry twenty-four thousand twenty-foot (TEU) containers.

Rose George recounts her voyage aboard the *Maersk Kendal* in her book *Ninety Percent of Everything*. When the book came out in 2013, *Kendal* was a midsize ship with its 6188 TEU capacity.[4] She mentions the excitement when the *Emma Maersk* was launched in 2005, with a capacity of 15,000 TEU and the anticipation with which the Triple-E class, which would carry 18,000 TEU, was expected by 2014.[5]

Some expect ships to grow larger still – even if it's becoming clear that the 'future profitability of ultra-large containerships,' a promise on which shipping companies and investors count, 'often turn[s] out to be a state-driven politics of scale' that is 'rarely fulfilled.'[6] Some even suggest that shipping companies regret the race for the biggest ship, as few ports can accommodate the vessels; even if ports can, absorbing the thousands of containers they disgorge can choke both ports and their hinterlands.

'How will we change all this?' asked Little Captain as we watched the ships that surrounded us in port. The task of turning shipping into a sustainable industry is immense, while the potential of a tiny ship like ours to make that happen seemed limited.

My own immediate concern was more the mysterious novel virus that had changed the world in our absence. We knew far less about the pandemic than did people ashore. We wore our masks and gloves – which were still a thing at that time – and liberally sprayed disinfectant on hands and surfaces. Despite the many stevedores, port officials, and the ship's agent visiting the *Avontuur*, compliance on board was very limited. I thought that Captain had not really stressed the importance of being careful to avoid infection well enough at the outset.

The risk of contracting the virus in Honduras was probably not that significant. According to the World Health Organization, there had been a total of three hundred and forty-three cases in Honduras by that time, with only thirty-one on the day we were in port.[7] Few, if any, were anywhere near the port district. Even so, I did not like the prospect of possibly being sick at sea with the limited medical care so typical of ships. Being stuck at sea is one thing. Being a floating vector of transmission is quite another.

The voyage from Elsfleth to Puerto Cortés had taken almost twelve weeks. But the slow and labour-intensive cargo operation made me realise all the more that, we're putting in an awful lot of work to transport a tiny amount of cargo.

Clean transport, dirty cargo?

On December 10, 2020, the Japanese shipping company Mitsui OSK Lines, known as MOL for short, announced a deal with Tohoku Electric Power to build a coal carrier equipped with a wind-powered propulsion system. Without a trace of irony, the company argued that this vessel would be the 'world's 1st coal carrier equipped with a "sail"' in a bid to reduce its greenhouse gas emissions.[8]

MOL's announcement reflects the mounting pressure on the shipping industry to curb its carbon emissions.[9] In this context, the decision to retrofit a coal carrier with sails illustrates the narrow focus of the shipping industry concerning climate change. Shipping, the industry adage goes, is a mere handmaiden of global trade. Demand for shipping is a 'derived demand' because demand for shipping is secondary to global demand for goods and commodities.[10] If there is demand for transport, ships will deliver. With this attitude, the shipping industry pleads innocence – or at the very least ignorance – of the environmental effects of the products they ship.

The government of Japan, where MOL is based, has since announced that it 'aims to achieve net-zero greenhouse gas emissions from international shipping by 2050.'[11] This comes after the country long 'opposed binding GHG targets, proposing instead a compromised set of "aspirational" targets (amounting to 50% decarbonisation by 2060) which would not to [sic] be in line with a carbon budget for the shipping sector that matches the goals of the Paris Agreement.'[12]

Japan has changed its tune and now supports net-zero-emission shipping and urges the IMO to revise its target upwards.[13] But what good does it do when they're equipping coal carriers with sails? It would, after all, be impossible for Australia and Indonesia to export coal if no shipping company carried it. Together, these two countries account for two-thirds of all global coal exports.[14] Cutting two-thirds of 1.2 billion tonnes in coal exports would reduce carbon emissions from both shipping and burning coal.

If the shipping industry is truly committed to climate action in line with the Paris Agreement, it should indeed consider not only *how* they ship, but also *what* they ship.

Coal exporters may need more than a nudge to change their ways. The 'Australian Way' to 'Deliver Net Zero' will do nothing of the sort. The government insists it can meet its commitments under the Paris Agreement, while 'it will not shut down coal or gas production or exports.'[15] They're not fooling anyone with such hollow words. They don't even try to conceal their disregard for the issue at hand.

In the more detailed version of the Australian government's 2021 vision, their reasoning becomes clear:

> While most of our major sectors will grow strongly to 2050, even as the world decarbonises, some sectors will face global headwinds. We will continue to export our traditional energy exports for as long as our customers demand them. If we were to withdraw supply and reduce our exports, other countries would fill the gap in supply. Australia's coal and gas export industries will continue through to 2050 and beyond, supporting jobs and regional communities.[16]

The 2021 Production Gap Report makes the underlying problem abundantly and unambiguously clear: 'governments' planned fossil fuel production remains dangerously out of sync with Paris Agreement limits.'[17]

Meeting the temperature goals of the Paris Agreement means leaving fossil fuels in the ground. And if exporters are intent on digging it all up, as Australia's government clearly is, perhaps the shipping industry should step up and refuse to carry any more coal. This is precisely what Eastern Pacific Shipping, a small bulk carrier based in Singapore, announced in its 2022 Environmental, Social & Governance policy.[18]

Shipping symbolises the collective action problem that is climate change better than almost any other industry. It happens mostly on the high seas. It is crucial to the functioning of the globalised just-in-time economy. In sum, every country needs it, but no country wants to take responsibility for it.

The underlying problem is that economies are organised through nation-states, whereas climate change doesn't stop at borders. The trade-off in our existential collective action problem is between gross domestic product and a liveable planet.

In 1972, researchers used excitingly novel and powerful computer technology to calculate that the global economy could not keep on growing indefinitely. There are, they argued *Limits to Growth*, in their report by that name, submitted to the Club of Rome.[19]

In 2008, Graham Turner published calculations of how the prognosis in the *Limits to Growth* maps onto observations between

1970 and 2000. Worryingly, he concludes that observations 'most closely match the simulated results of the Limits to Growth "standard run" scenario for almost all the outputs reported; this scenario results in global collapse before the middle of this century.'[20]

In 2021, Gaya Herrington updated the calculations once more, only to find that the 'two scenarios aligning most closely with observed data indicate a halt in welfare, food, and industrial production over the next decade or so, which puts into question the suitability of continuous economic growth as humanity's goal in the twenty-first century.'[21]

Herrington works for KPMG, hardly a club of deep-green environmental activists. They are the foot soldiers of global capitalism, who are gearing up to make money from the apocalypse, as they have from helping corporations destroy the planet.

Even so, Herrington's message is clear: continued economic growth throughout the twenty-first century is highly unlikely. This is not a matter of people voluntarily opting to consume less, as it's unlikely that enough people will choose that option.

'But that's not the choice,' Andrew Curry argues. 'The choice is that either we end up with unmanaged decline, which would be catastrophic, or a managed levelling out of our economies, shaped by a shift in social values and expectations. We need some politicians who are willing to be honest about this.'[22]

And yet, many keep on treating 'net-zero' as a reasonable and feasible trajectory towards a liveable planet, by pursuing action in perfectly isolated silos. We might, for example, decarbonise shipping without thinking about the kind of economy the shipping industry enables. Even Ikea and Amazon have now committed to shipping their goods with zero carbon ships by 2040.[23] How they ship is important, but it's only a small part of the challenge we face.

'Continuous material growth,' Vaclav Smil reminds us, 'based on ever greater extraction of the Earth's inorganic and organic resources and on increased degradation of the biosphere's finite stocks and services, is impossible. Dematerialization – doing more with less – cannot remove this constraint.'[24] Can the shipping industry really claim to play no part in what exactly is shipped? I would argue it can't.

Following this logic, MOL's efforts to 'decarbonise' shipping would be the solution. It brings down the carbon intensity of the vessel, which helps the company meet IMO regulations while building its credentials as a company committed to reducing its carbon footprint. Shipping decarbonisation is essential, but so is rethinking the fossil fuel exports and unsustainable levels of consumption it enables. Without ships, Australia would not be able to export coal.

'Time flies,' I wrote in my notes while at sea. 'The days just disappear.' With every day, drastic climate action is delayed, our remaining carbon budget shrinks. As does the chance of keeping global warming below 1.5°C.

Paradise denied

After leaving Puerto Cortés on April 11, we slowly made our way to Belize. With Easter ahead of us, we were not expected in the port of Belize City until April 16. So, we spent a long weekend anchored off Goff's Caye.

Goff's Caye is a private island, between the Belizean coast and the Turneffe Atoll. Even though the tiny plot of sand has no permanent inhabitants and housed no tourists at the time, owing to the pandemic, we were not allowed to step ashore. So, instead of sipping rum on the beach, we were left marooned at anchor on our own island.

Forbidden from going ashore, we mended sails and swam. We finally stepped out of our watch system that dictated our days by rigid clock time, four hours at a time. Despite this downtime, or perhaps due to it, simmering conflicts started surfacing. The Red Watch, Cook argued behind our backs, were nothing but slackers who got a free pass because they were the Captain's

Goff's Caye, Belize, seen from the *Avontuur*, April 15, 2020

watch. Martin, used to the work and discipline of the *Bundeswehr*, concurred. To me, it looked like little more than a power play between Cook and Captain. She wanted to undermine his authority by pitting us against each other. And he did not want to give her the pleasure of acknowledging how successful she was at this.

In private conversations, Captain reflected on how most of these tensions emerged because of power struggles inherent to social life. While we had voluntarily joined the ship and willingly subjected our time and freedom to the rules of the ship under the authority of the captain, no one planned to be locked out of land on this

The *Avontuur*, seen from the water while swimming, April 14, 2020, off the Belizean coast

ship, without any shore leave. By this point, we had entered a wholly new stage of the voyage, which extended beyond our plans and mental preparations. As we could no longer leave as foreseen or make any plan of our own volition, the lack of autonomy started to sink in. Not being able to step foot ashore the nearby tropical island was a constant reminder of our predicament.

Cook responded to this by strengthening her circle of acolytes. Little Captain grabbed every opportunity to assert his imagined superiority. Neither of them had any real power; but neither did they wholly submit their autonomy to the chain of command on the ship.

The Red Watch bore the brunt of this. My own embrace of idleness was still not universally shared. The work we did while at anchor was necessary, but simultaneously served a larger social purpose. It made us into a crew of sailors. It created a shipboard subculture characterised by work and effort, a dedication to the principles of sail cargo, an eagerness to learn the craft and skill involved in sailing and maintenance, and a commitment to the common good through a negation of the self.

'I look after this ship well. It's my home,' Bo'sun told Peggy, who'd expressed worry about the outstanding maintenance of the vessel. 'It won't deteriorate on my watch. But I'm not rushing my work. I can't finish things too quickly, as I need to justify my paycheck.'

Of course, he was largely joking. The pace of repairs is indeed slow, but constant. Committed and careful, but never frantic. I greatly enjoyed mending sails and spent much of my time doing exactly that – though Benji and I were so slow that we might as well have been sipping drinks on the panorama deck, as we'd started calling the spacious cover of the cargo hold. If anything, our jocularity right next to Bo'sun, who was mending sails with a pipe dangling from the corner of his mouth, sped him up.

While at anchor near the coast, we were able to access the internet through Captain's *Skyroam* connection, which conveniently connected us to the world whenever we were in reach of shore. If no more than three people used this at the same time, we were able to use messaging services, email, and make the occasional call. While welcome, I'd started noticing how, whenever we had connection, the atmosphere aboard changed.

On April 16, 2020, the forty-four-metre-long schooner celebrated her hundredth anniversary. By this point we knew that there would be no shore leave in Mexico, our last cargo port, leaving us aboard until returning to Germany. Built in 1920 at the Smit shipyard in Stadskanaal, Groningen, the *Avontuur* first went to sea

Bo'sun and Jennifer mending sail, April 20, 2020

in the same year as Captain Alan Villiers, who went on to become the self-styled chronicler of the last days of sail.[25]

When the *Avontuur* was built, the shipping industry was rapidly carbonising the last remaining vestiges of wind-propelled cargo transport. It took well over a hundred years, between the first commercial steamship in 1807 and the 'last grain race' in the late 1930s, for fossil fuels to push out sail entirely.[26]

In 1920, world maritime trade was growing. In 1887, total seaborne trade added up to one hundred and thirty-seven million tonnes, which increased fourfold to just over half a billion tonnes by 1950.[27] The real expansion was yet to start. By 2019, the industry had increased twenty fold, to eleven billion tonnes of cargo traded annually.[28] Even at the time she was built, it was clear that the *Avontuur*, with her small hold and one hundred and twenty-four tonne capacity, was never going to be suitable for transatlantic trade.

Villiers was right, sail was fast disappearing from the world of commercial shipping. The *Avontuur* was, even when launched, a relic. The First World War left my native Flanders in ruins. But when the war ended, the massively expanded military fleet boosted Europe's capacity to trade under mechanical propulsion. The necessary infrastructure to maintain and refuel coal- and oil-fired ships was at its height, in no small part thanks to the war.

To mark the hundredth anniversary of the *Avontuur*, we dressed up in whatever clothes we had on board that could make us look like early twentieth-century sailors, so we could send Cornelius a memento of the centenary. We celebrated every anniversary aboard – the *Avontuur*'s was no exception. This was a particularly cheerful moment on our voyage. We were anchored in quiet seas with nothing much to do but wait for Easter to pass. So, we spent far more time together as a crew than we had so far. While at sea, you see your watch members all the time and those in other watches barely at all. Generally, our social life scarcely extended beyond our respective watches, so it was a welcome change to have our 'contact bubble' grow.

Despite the cheerful mood during these celebrations, the atmosphere aboard had shifted from a holiday camp for adults to a real community of people bound by circumstance. We were no longer willing to downplay our differences. We started magnifying minor events and inevitable character flaws. Ashore, we would never have stuck together for this long without open conflict. Thanks to being at sea, though, we managed to retain a semblance of harmony.

While we were still at anchor, Captain briefed us on the practical challenges we faced in Belize City. 'Absolutely nobody' would be allowed ashore, 'not even for medical emergencies. In case of an emergency, the port would be closed to us, and we would have to search for another option.' After loading cargo, we'd head 'to Mexico a.s.a.p. because the shops are still open.' We'd provision there until our planned stop in Horta. The possibility of re-provisioning at Freeport in the Bahamas was uncertain, so we'd be wise not to count on it.

'The world is a magic show,' he continued. 'You can choose to be entertained by it, or you can be informed by it. You can look at the magician and be amazed by his or her tricks, or you can look at the bigger picture and see what's really happening. It's not that I dispute the severity or the seriousness of the coronavirus pandemic. There is suffering. There's death. There's heartache.'

Indeed, at that time, my father was in hospital struggling to recover from COVID-19.

'When the dust settles on COVID-19,' he said, 'the world will be forever changed. Many industries will suffer. But a handful of other companies are already thriving and will continue to do so. And for well-informed investors, it will be the best investing opportunity of this decade … and perhaps of our lives.'

For us, on board the *Avontuur*, the implications were a lot more mundane in the first instance.

'We are free to have group meetings larger than two,' Captain reminded us. 'We can meet with non-relatives and non-family members. We don't get heavy fines for doing so. We only have to wear masks and gloves for protection in ports and will do so more strict [sic] from here onward. We still have travel and have a purpose. The world will adjust on every level. Hard times bring people together.

'The process of uniting means for us we are required to let go of resentments in order to find a common ground beyond carrying cargo under sail,' he concluded in an attempt to bring this all back to how we'd have to live together for months to come. 'Don't expect anything in return when giving freely from the heart.'

Just before heading into Belize City to load four tonnes of cacao beans, we sent our order for personal goods to Cornelius. For lack of shore leave, we had to place our personal orders through the ship chandler. What we needed most? Chocolate.

Tonnes of coffee

Thirty tonnes. Nearly half of the *Avontuur*'s hold. That's how much Timbercoast's biggest customer shipped in 2020. Teikei Coffee is a Hamburg-based company that offers coffee subscriptions. When signing up for a twelve-month subscription, you commit to buying at least four kilos of coffee at thirty euros each. More frequent deliveries, six or twelve times a year, are also possible, as are large quantities for companies or co-housing communities.

This is how my fellow shipmate Pinkie learned about Timbercoast, in her *Wohngemeinschaft* (or housing community) of thirty-five people in Bern, Switzerland which ordinarily buys organic locally grown food in bulk. As coffee can't be grown within the country, it posed a challenge for the group. How to get coffee in a way that is socially fair and environmentally responsible? Teikei offered the solution.

'The Teikei project is a CSA, community supported agriculture,' its founder Hermann Pohlmann told me. 'That means we will need farmers, who really want to understand, who want to go in the direction of a different economy. We don't speak any longer about the kilo price and the amount of coffee, we speak about "Farmer, what do you need and what can you give us? How many people can you produce coffee for? So how many families can we include?"'

Teikei's partner farmers in Mexico knew about the *Tres Hombres*, sailing cargo across the Atlantic. They suggested Pohlmann look into this option; they thought a fair and transparent cooperation ought to incorporate environmental awareness too.

For the *Tres Hombres*, Mexico was too far to be practicable. And Teikei's initial half tonne order in 2017 meant that the *Avontuur* was not a viable option for Cornelius, either. Port costs would be painfully high for such a small amount. The detour into the Gulf of Mexico would take too long, making the voyage too costly.

Over the following years, Teikei's orders grew from ten tonnes in 2018, to twenty in 2019. These amounts made worth the trip to Veracruz. When Pinkie and I were aboard, we loaded Teikei's largest order yet: thirty tonnes of green coffee beans from Oaxaca.

Pinkie had already given up her job at the Green Party of Switzerland before she decided to join the *Avontuur*. Travel beckoned. The *Avontuur* offered the possibility to cross the Atlantic without flying, while also learning how to sail. This seemed far more appealing than the risk of ending up with an 'old creepy guy' on a yacht. Crew-finding websites are not always reassuring for women travellers.

'I think it's very chaotic and I like it,' she summed up the experience on board. 'You know, it suits me very much.' Life on board did, indeed, feel rather more chaotic than Timbercoast's neatly crafted voyage charts and social media suggest.

On board we knew precious little about what we were actually going to ship. It was only when I spoke to Hermann after disembarking that I fully understood his unique way of trading coffee.

'We don't work any longer with the prices that are made on Wall Street,' he stressed, 'which has nothing to do with the work, what the farmers are doing.' Teikei's pricing model is based on cost calculations, rather than on market forces driven by the daily prices captured by the International Coffee Organization, or what the average consumer is willing to pay in a supermarket.

Teikei's pricing builds on a fully transparent cost calculation listed on their website. This 'calculation results in a cumulative total price of €31,64 per kilogram of coffee.' They're charging subscribers thirty euros per kilo. This includes seven euros per kilo for the purchase of green coffee in Mexico, four euros for

Stevedores loading a cargo of green coffee in Veracruz, for Teikei, April 27, 2020

sail transport across the Atlantic, six euros for Teikei running costs, about five euros for roasting and the weight loss incurred in the process. Then there are minor charges for packaging, taxes, insurance, ground transport, and building some 'development and security reserves' for Teikei.[29]

This makes for expensive coffee. But not exceptionally so.

Market Lane Coffee, one of Melbourne's favourite roasters, charge on average twenty Australian dollars for 250 grams, or over fifty euros per kilo.[30] While they commit to transparency, and claim to pay 'sustainable prices,' they're not actually listing any details on their website – as Teikei does.[31]

Teikei does not simply buy coffee. It builds long-term relationships with producers in one valley. Hermann explains they 'want to make a change in this valley with the whole food system.' And sailing the coffee across the Atlantic Ocean is part of that, as their website highlights:

> Together with the Timbercoast shipping company, we represent Mission Zero. The aim of this mission is to transport consumer goods in an environmentally friendly way over land and water in order to reduce emissions along the entire value chain. By transporting goods by sailing ship, pollution of the oceans by heavy oil and noise is avoided. In addition to transporting goods for partner organisations, the company is financed by enabling adventurous people to sail on board and by investors and shareholders who want to support the project.[32]

The problem with this partnership is that it has its limits, because the *Avontuur* has a finite loading capacity, while Teikei is growing.

'We are growing very fast,' said Hermann. 'If we don't have enough space to bring the coffee by sailing, we would use container ships. For us, the more important thing is that farmers can grow, and the farmers can get a good price, get a better life. If there are sailing ships, of course we will use them. If there are bicycles, we will use bicycles.'

Sail cargo companies plying the Atlantic have to find west-bound cargo. If they don't, companies like Teikei effectively pay for the entire round trip, rather than just the one-way, Europe-bound leg their cargo involves. As they generally can't find cargo headed west, they now pay four euros per kilo for shipping, which Hermann thinks is 'much too much.'

Cornelius told me that the 2020 price he charged for transport on *Avontuur* was €4.20 per kilo, or €4200 per tonne. This does not include port charges, which are borne by cargo owners. Fairtransport, the company behind the *Tres Hombres* and the *Nordlys*, declined to share their prices with me. (Though I did find out they charge roughly fifty eurocents per tonne-mile.)

For comparison, the *Freight Rate Index* – the average cost for the shipment of a forty-foot container – stood at US$1500 in May 2020, peaking around US$11,000 in September 2021.[33] At the time of writing, in May 2022, the rate for a shipment from Veracruz to Hamburg is roughly US$3500.[34] With a loading capacity of some twenty tonnes of coffee per forty-foot container, transporting 65 tonnes of cargo on a container ship would cost somewhere between US$4500 and US$33,000. This is roughly ten to sixty times more than the same cargo load on the *Avontuur*, which was priced at €273,000.

Teikei were not the only company with cargo on board. Other cargo owners included the Leipzig-based Café Chavalo and the Italian-Austrian company Brigantes, who are restoring their own vessel in the Sicilian town of Trapani. But what will happen when newer and larger ships offer transport under sail as well, likely at far lower prices?

Guillaume le Grand, who runs the trading company Trans-Oceanic Wind Transport from Le Havre, plans to reduce prices with the ship they have started to build. He claims they will be able to offer freight rates that are a tenth of those charged by Timbercoast in 2020. The Canadian-Costa Rican company Sail Cargo Inc. also plans lower rates than the *Avontuur* and the *Tres Hombres* once their flagship *Ceiba* sets sail. In the meantime, the going rate will likely remain somewhere between three and five euros per kilo of freight, for a transatlantic crossing. This generates real potential for companies like Teikei to invest in their own ship.

Many of Teikei's customers first subscribed to their coffee deliveries because the cargo was sailed across the ocean; many stay on because they find the locally embedded CSA model more trustworthy than the fair-trade or organic coffee found in supermarkets. But how much can sailed, fair-trade, organic coffee from community-supported agriculture do to decarbonise the shipping industry? Even in the UK, where fair-trade coffee is commonplace, only one in every four cups uses it.[35] If, after decades of mainstreaming fair-trade coffee, it remains the exception, what chance is there that everyone will be happy to pay a hefty premium on sustainably shipped coffee?

Even if every last bean of coffee were shipped under sail, this would not make a dent in the global shipping industry. The total trade in coffee, 166 million bags weighing sixty kilos each, add up to just under ten *million* tonnes.[36] Compare this to the eleven *billion* tonnes of cargo transported per year. A trillion cups of coffee, give or take a few, make up only one-tenth of a percent of all maritime cargo.

Luxury or necessity?

After seven weeks at sea, most of them in the tropics, I finally regained some dignity by washing my bedding. Ordinarily, we'd take all sheets, towels, and clothes for a machine wash in port. While I kept my clothes clean, washing them by hand on deck, the prospect of having to wash my sheets that way was simply not appealing. On reflection, neither was *not* washing them.

The ocean, despite its abundance of water, is not an easy place to stay clean. That feeling of permanent humidity after going for a refreshing ocean swim? It's the salt sticking to your skin, absorbing humidity from the air. Soap not lathering properly? It's the salt messing up the reaction between water and soap. Bedsheets permanently damp? It's your own sweat that eventually makes for salty sheets that never feel dry. All this gets worse in the tropics.

Given our limited supply of fresh water, which used diesel to produce, we washed everything with seawater. The dishes, our clothes, ourselves. Only after washing and rinsing, did we rinse with fresh water. Being at sea made me reassess my needs. Clean sheets are high up my list.

The world, even while we were stuck at sea during the pandemic, is, of course, a lot bigger than the *Avontuur*. And tackling the climate crisis is something I cannot do on my own. But our microscopic predicament made me wonder: what exactly is *necessary*?

Is it necessary to ship Atlantic cod halfway around the world to China, so it can be filleted more cheaply, only to ship it all the way back to Scottish supermarkets?[37] Is it necessary to ship recycling halfway across the world for processing, just because it's cheaper than processing it at home?[38] Is it necessary to have sweatshop labourers halfway across the world break their backs so we can buy a single-use Halloween outfit?

These may be rhetorical questions, but these things actually happen. As the shipping industry is expected to grow threefold

Loading a cargo of cacao in Belize, April 17, 2020

by 2050, it's worth asking precisely what we ship.[39] Shipping, let's remember, has already grown twenty fold since 1950.[40]

In 1950, Earth's population of 2.5 billion people shipped a grand total of half a billion tonnes, or two hundred kilograms per person in a year. By 1970, the population had increased to 3.7 billion, while the amount of cargo had increased five fold to 2.5 billion tonnes, or 675 kilos per person. Cargo transport thus increased three times as fast as the population. In 2019, we were 7.7 billion people on the planet, shipping a total of eleven billion tonnes, more than 1.4 tonnes per person. While the world's population is tipped to stabilise around eleven billion by the end of the century, so many people are yet to start consuming the way I, and probably you, do.

This is not to say that the upward trajectory of cargo volumes is unstoppable. Indeed, the data scientist Hans Rosling warns that it is a fallacy to assume that an upward trajectory will continue.[41]

As mentioned earlier, the material limits of the planet will likely slow down, or even halt, economic growth in a matter of decades.[42] If we do not plan for this, having to live within planetary boundaries will only become more difficult and painful.

Changes in the way we organise the world will likely impact what and how much we ship. Fossil fuels, for example, account for about 40 percent of all maritime cargo.[43] While climate activist Bill McKibben thinks that is good news – because eliminating fossil fuels would immediately also cut shipping demand – the real impacts won't be as clear-cut.[44] Strong climate action that significantly reduces reliance on fossil fuels *could* lower demand for shipping.[45] In other words, if we cut fossil fuel use and organise the shipping industry more efficiently, we would not only cut emissions from *not burning* fossil fuels, but we'd also stretch our remaining carbon budget by *not shipping* those fuels in the first place. Shifting to biofuels and e-fuels would likely increase demand for shipping again, as production and consumption locations won't align very neatly either.

The major logistical issues in 2021, resulting from a lack of empty containers in export countries and sky-high spot rates, prompted many countries and industries to rethink their supply chains.

A part of that solution may be shifting to a circular economy, which would use resources in a smarter and more sparing manner. This, when properly incorporating transport into circular planning processes, could allow us to produce more with fewer resources and less transport.

Simply re-organising the shipping industry could save up to 38 percent of its emissions.[46] That is not from shipping less, just from organising supply chains more efficiently.

Shipping tiny amounts of cacao and coffee may raise awareness about the scale of the industry, though the *Ever Given* blocking the Suez Canal in 2021 made global news headlines in a way that sail cargo initiatives have not.

We have become a world of consumers, though acting solely through consumption is an admission of defeat. Our individual consumption habits may make us feel good about ourselves, or 'eco-pious' as Sarah McFarland Taylor calls it, but they won't change the world.[47]

In Tenerife, I stocked up more than enough for a three-week voyage. I mostly bought Vivani chocolate at *Las Hierbas Salvajes* (*Wild Herbs*), a tiny organic food shop on Calle Ramón y Cajal, beside the Barrance de Santos, an intermittent stream carrying storm water from the hills rising between Santa Cruz and the university town of La Laguna.

Tres Hombres's chocolate costs €3.75 for an 85-gram bar, which adds up to €44 per kilo. *Avontuur* chocolate, on the other hand, costs €6.50 per bar of 58 grams, which translates to €112 per kilo.

No sailed chocolate was included in our rations, and neither was it available for purchase on board. It would not have been cheap, but one doesn't bargain when it comes to such necessities.

Unfortunately, the challenge is not to decarbonise a few luxury 'necessities.' We need to decarbonise all our necessities. This is why Bill Gates opens his 2021 book, *How to Avoid a Climate Disaster*, by asking how we get from 52 billion tonnes (or gigatonnes) of greenhouse gas emissions a year down to zero.

The latest Intergovernmental Panel on Climate Change estimates that the remaining global greenhouse gas budget is running out very rapidly. To have a 67 percent chance of keeping global warming below 1.5 degrees, our remaining cumulative carbon budget is 230 gigatonnes of carbon dioxide. At current emissions levels of around 50 gigatonnes per year the remaining carbon budget will be depleted in less than five years. The riskier option of having only a one in three chance of keeping temperature increases below 1.5 degrees, gives us more time.[48] By the time you read this, the budget will have shrunk further.

The longer we wait, the less 'carbon budget' we'll have left, and the faster we'll have to act. That's why it's so frustrating that the IMO has dragged its feet for so long. We don't simply need to get emissions down to zero, we need to stop emitting before the remaining carbon budget runs out. At current emissions levels, that will be well before 2050.[49]

Climate change, the environmental philosopher Timothy Morten argues, is a 'hyperobject' – an issue so large and complex that it is almost impossible to grasp.[50] The change is minute and gradual, and therefore invisible. But planetary tipping points, once crossed, may lead to irreparable damage. The shipping industry, like climate change itself, is a huge and complex beast. It is a 'hyperobject' too. The world as we know it cannot do without sea freight. Supply chains – and livelihoods – across the planet would collapse if ships stopped. But the world cannot allow the industry to continue its business-as-usual. At the risk of repeating myself: shipping's contribution to carbon emissions is huge. Shipping itself is a necessity, though far from everything we ship is.

Unraveling

On April 18, somewhere halfway between Belize and Veracruz, Jen, Benji, and I started looking into the nuances of sail-handling, by dissecting the fine detail of trimming and manoeuvres during our watch. Rather than doing this in the midst of manoeuvres, we wanted to do this quietly when we had the time to consider the precise functioning of the ship. The First Officer often joined us. His knowledge of sailing had improved greatly, though he remained somewhat flummoxed by the intricacies of sail. We took our time discussing the ship's lines and their functions. Together, we improved our sailing craft. Rather than fantasising about our futures ashore, likely months away, we tried to make the best of our time at sea, though none of us seemed to find that particularly easy.

Cook made it clear to all of us that she was fed up with it all. She wanted the voyage to end. In the tropics, fruit and vegetables spoil fast. Especially if they were refrigerated before entering our non-refrigerated dry store. Our fridges, two large household ones, were simply not big enough to hold all our fresh food. None of this made her life any easier. Between leaving Belize City and arriving in Veracruz, we had serious amounts of fresh fruit to get through. But rather than eating it while fresh, we were going through the pretty-much-rotten food from the dry store every day.

'We might as well eat fresh food while it's fresh,' Benji and I suggested, because we expected to get fresh food in Veracruz.

'How do you know for sure this will happen?' she snapped. 'No provisions are certain until we have them aboard.'

Never mind that we had provisioned at three out of three ports of call, despite lockdowns. Never mind that Captain stressed several times that we would not leave Mexico without ample fresh food.

After yet another confrontation with Benji, Little Captain snapped at Jennifer, saying he was not used to 'working with assholes,' but that he was 'trying to make the team work.'

Just before starting our watch, Pinkie told me she'd had a bad day. She was increasingly frustrated with everything and everyone. To her, we had too much time on our hands and too little to do.

On April 23, Cook and I had a run-in after lunch. I honestly can't remember what it was about. But that barely matters. I snapped at her. It was so bad that nobody seemed to dare enter the galley until I was done. I, too, was well and truly fed up with it all.

That evening, as we were about to reef the mainsail, we ended up taking it in altogether. Rather than the wind picking up, it suddenly died. As the *Avontuur* started rolling for lack of wind in the sails, Captain commanded us to take in the schooner gaff topsail.[51] As we tried to douse the sail, the sheet bend came loose. Earlier in the day, someone had practised tying that knot. A little later, they put the knot through the clew and we hoisted the sail. There was only one clew where we used a sheet bend, so we hardly ever tied that knot. For whatever reason, with the clew loose, the sail got stuck around the spreader. As Bo'sun tried to douse the sail, the ship kept rolling, the schooner kept on flapping. Within half an hour, the schooner gaff tops'l was neatly folded on deck, though not without breaking the VHF transmitter. After we debriefed on the aft deck, just as we thought the worst was behind us, the schooner ripped horizontally through the middle. This made one of the most important sails aboard – second only to the staysail – utterly useless. Even when the wind picked up again, we'd have a hard time gaining speed.

'Further to the schooner sail,' Cornelius reassured our family and friends through one of the many emails he sent them throughout the voyage, 'the AIS antenna got damaged when the fore topsail was doused. That's why we are not receiving *Avontuur*'s position anymore.'

On one of the last nights before arriving in Veracruz, I was in the galley with Jennifer and Benji during the evening watch. Jen asked where Michael was.

'Michael is speaking to the Captain,' said Benji, as if that was the sensible response.

Michael, however, *was* the Captain.

At that moment, it dawned on me how challenging it must be to act as a ship's Captain. You are responsible. You are in charge. You are never off duty. And you are hardly ever alone. As a result, the crew confide in you, lean on you, trust you. But you're effectively on your own. By that point, we'd been aboard uninterrupted for nearly two months, without any shore leave. The closest we came was a couple of days at anchor off Belize. No wonder Michael was speaking to the Captain.

The following evening, on April 24, Little Captain called our watch to discuss the tensions we faced – to complain about Benji, really. During our evening watch, between 20:00 and midnight, we listened to his frustrations. Captain actively stayed out of the conversation. He overheard it from the chart room, but did not engage. He did little to address Little Captain's frustration, either.

Captain regularly asked Joni, the other deckhand, to step up, whenever Bo'sun was resting below decks. Captain would even call in Joni when Little Captain, the other deckhand, was on watch. Given Captain's lack of trust in Little Captain, this may have been a sensible thing to do, but it rubbed salt in the wound. Whatever the reasons, Little Captain regularly clashed with Benji, who would stubbornly defy him. Benji knew all too well that taunting Little Captain would not lead to anything good, but he could not help himself. Two men who were used to getting their way were not about to budge.

Cornelius had promised Little Captain greater responsibility. But Cornelius was not on board as the master of the ship. Little Captain may have had a good connection with Cornelius, but he and Captain actively disliked each other. Even so, signing on as deckhand is asking for trouble if you deem yourself a master. I would no longer join a project as a research assistant, as I consider myself an independent researcher. I would not cope very well in such a subservient role. When stuck at sea for far longer than most of us had bargained for, such tensions were hard to rationalise away.

We may have ripped the schooner, but we had damaged a great deal more by the time we reached Veracruz. Each one of us had imagined 2020 to be different and we were powerless to change the cards we'd been dealt. We all reacted very differently to the frustrations of being stuck on the *Avontuur*. If Mexican authorities had allowed us shore leave, quite a few of us would have likely jumped ship.

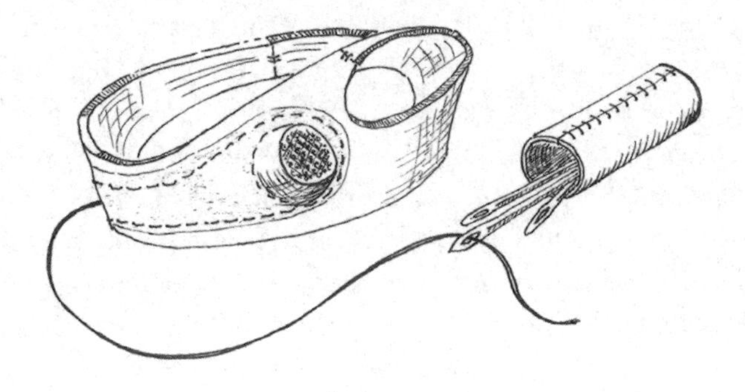

5

Point of return

Veracruz

By arriving in Mexico, we reached the furthest point on our voyage. The point of return. We all needed this, but coming alongside in the port of Veracruz was nowhere near as easy as it had been thus far in other ports. Cornelius reported, to those ashore, that we had made it:

> *Avontuur* has dropped anchor on Veracruz anchorage 25.04.2020 at 2318h local time and will proceed inwards alongside pier 4 today at 0730LT.

If only it was that easy. The next day, a Sunday morning, we woke to an unexpected storm. It was neither forecast nor reported by the port authority. The wind was blowing at Force 9, gusting 10.

For all the technological advances since the *Avontuur* was launched in 1920, the ship had no anemometer – a device to measure wind speeds by means of four half spheres that turn an axle – named after *Ἄνεμοι*, or Anemoi, the Greek gods of wind. We relied wholly on how Captain read the sea and the wind. While the GPS gave us an accurate sense of our Speed Over Ground, we had to calculate the wind speed from the sea state, making an estimate of the speed based on the Beaufort scale.

Despite having dropped both anchors, the ship was dragging. In the middle of the storm, a car carrier anchored one nautical mile behind us radioed to request we move. We were sliding closer to her at a speed of 1.5 knots. A collision seemed imminent.

Port authorities did not allow anyone to come into port during the storm. In fact, they asked ships to leave and ride out the rough weather in the Gulf of Mexico.

It was clear we could not leave. Our engine lacked the power to keep the ship in position, even with both anchors down. Once the storm petered out, before we hit the ship anchored downwind, we waited for a call from the port to tell us a berth was available, which could take a while. We'd planned to quickly pick up cargo and head back across the Atlantic. Cornelius was keen for us to get to Hamburg. But with the schooner sail ripped and our VHF transmitter broken, we could not leave before all the repairs were finished.

In the afternoon, Captain asked what we hoped to do on the return voyage. I proposed focusing on learning by doing, building an understanding of ship manoeuvres so we could take command. I wanted to *do* more on the return voyage, because by the time we arrived in Veracruz, the social fabric on board was in tatters. Everyone was exhausted, irritated, and tense. In port we wouldn't be allowed to get away from each other even for a single day. COVID-19 still held the world in its grip. Tensions between watches had become untenable. The idea that the Red Watch was somehow Captain's favourite dampened enthusiasm for sailing in the same constellation.

Captain announced that in port we'd shift to a Day Watch schedule, where everyone was expected on deck between breakfast and dinner. Only those who volunteered to take the nightly Harbour Watch would stay up at night and sleep during the day. On the return voyage, we'd all be in new watches. My spirits sank. Our Red Watch had been an absolute pleasure. Benji, Jennifer, and I got along famously. And we had the pleasure of getting to know Captain better than any other watch. Despite the tensions with Little Captain, we made a great team.

After months at sea, time in port was both a relief and a source of frustration. We were not allowed to leave the grimy port, so we spent a week bathing in soot particles, dust, and floodlights that burned throughout the night. Veracruz, more than any other port we'd previously called at, confronted us with shipping pollution at a scale we had not seen before. It showed just how much investment ports need to make as they prepare for a 'clean' shipping industry.

I started reading *The Ends of the World* by the Brazilian philosopher Déborah Danowski and the anthropologist Eduardo Viveiros de Castro.[1] Its strong focus on life ashore, on *land*, struck me, much like Bruno Latour's *Down to Earth*, which I read midway across the Atlantic.[2] They all speak of terra, territory, Gaia, and Earth. But what about Ωκεανος, or Okeanos, the Titan son of Gaia and Uranus? Perhaps best known from the Trevi Fountain in Rome, Okeanos was the Greek god of the river Okeanos, which surrounded the flat Earth. Let's forgive them that misconception; 71 percent of the planet's surface is water. And oceans are where climate change is taking place most significantly.

Curbing emissions in port requires ships to use shore-based, low-emissions energy. Port owners must rethink infrastructure to allow the delivery of alternative, less polluting fuels to ships. This falls beyond the remit of the IMO, but requires regulation, concerted action, and massive investment – by port authorities and owners. UMAS, a London-based consulting firm in the maritime sector, calculated that 87 percent of transition costs would be ashore, while only 13 percent of investments would need to be for ships.[3] It's a challenge over which neither small sail-cargo companies nor major shipping companies like Mærsk have much influence.

Since 2015, the European Union has supported a group of small commercial ports along the North Sea coast to explore, develop, and trial structural ways to 'green' their operations. The ports of Oostende (Belgium), Vordingborg, Skagen, and Hvide Sande (Denmark), and Zwolle (the Netherlands) collaborate to reduce their environmental footprint by improving port operations and administration in ways that 'promote responsible growth' and support 'eco-innovation oriented development.' By focusing on small ports, it is far easier and cheaper to experiment, in the hope of finding solutions that could work in large ports too, though the initiative also aims to revitalise these small ports, which could help them relieve pressure from congested ports elsewhere in Europe. Much like small-scale sail-cargo initiatives, these are welcome interventions. But when looking at the scale of Veracruz, a tiny port compared to Shanghai, Tianjin, Guangzhou, Long Beach, Immingham, Rotterdam, Jebel Ali, or Singapore, I wondered: will these small-scale efforts be successful and scalable?

Labour Day

On May 1, 2020, International Labour Day, a local company picked up the schooner sail for repair. Not a sailmaker, as there was none in Mexico's third-biggest port who could work with the thick plastic material used for the *Avontuur*'s sails. It was a company that makes and repairs truck tarps.

Later that day, all ships sounded their horns, in an attempt to render the invisible fate of stranded seafarers audible. Our protest call came in response to an invitation from the harbourmaster, Gabriel Ángel Carreón Pérez. The day before, he circulated a message to all vessels in port, asking everyone to sound their horns at noon as part of an international appeal for recognition. It was then that it properly sank in. We were not alone. Many, if not all, seafarers were stuck at sea, marooned at work by COVID-19 measures. Many had been stuck aboard for far longer than we had.

The 2006 Maritime Labour Convention (MLC) stipulates that labour contracts for seafarers cannot exceed eleven months. Maritime workers can be asked to toil up to twelve hours a day, every day of the week, for nearly a year.

As the much-anticipated end of a contract approaches, any delay causes difficulties and frustration. But delays in tandem with the uncertainty of a global pandemic made life miserable for many workers at sea. Seafarers had little choice but to continue working. Putting down work and going home simply wasn't an option. At sea, there is no escape. In port, you might jump ship, but the captain holds your passport and it's unlikely you could leave the port compound without being stopped. Even if you could, it would mean the end of your working life in shipping, as you'd quickly get blacklisted by the agencies that most seafarers rely on for work. The options for officers are more plentiful, but the more numerous 'ordinary seafarers' have little choice but to stay put and agree to contract extensions.

The MLC is not perfect. It merely cements the exceptional nature of work at sea, so working hours and weeks far exceed typical working hours in the West. But the MLC provides enforceable standards for the many seafarers who work on ships that operate under flags of convenience issued by 'flag states,' which often lack strict labour, safety, and environmental regulations.

Force majeure is one of the few things that can override the Convention. A deadly pandemic fits the bill. When virtually every country closed its borders in March 2020, seafarers and the International Transport Workers' Federation (ITF) accepted contract extensions without much ado. They recognised that people ashore relied on the goods they supply. Seafarers are essential to keeping the global economy going. Even so, few countries have deemed maritime transport workers 'essential' or 'key' workers – something the International Labour Organisation (ILO) had called for since late March 2020. Such a designation would allow workers at sea to travel between work and home during COVID-19 restrictions.[4]

The ITF started sounding the alarm as early as April 1, 2020.[5] When the Philippines and Ukraine – two countries with large maritime labour forces – closed their borders in March that year, crew changes became practically impossible. Flag states – countries with 'open' shipping registries and large commercial fleets – allowed employment contracts to extend well beyond the eleven-month statutory maximum under the MLC. Up to seventeen months. That is nearly a year and a half of working at sea, often without a break. It took Panama – a major flag of convenience – until September 14, 2020 to repeal its decision on contracts and insist they would return to limits set out in the Convention.

As the 'crew change crisis' unfolded, major shipping organisations and companies, including Wilhelmsen, provided detailed day-to-day updates on which ports were allowing crew changes and under which conditions.[6] These regulations changed as we sailed.

I did not feel the same uncertainty that many fellow seafarers faced, and continued to experience two years into the pandemic. Stepping off anywhere in the Americas was not an option. But as our voyage was to end in Hamburg, that's where my journey would end. That much was certain. What wasn't so clear was just how long it would take us to get there, as is inherent to any sailing voyage.

The crew change crisis worsened as we travelled on, even though crew change became possible in a growing number of ports. This crisis resulted in what the ILO classifies as 'forced labour.' Brandt Wagner, Head of ILO's Transport and Maritime Unit, and Jan de Boer, an IMO representative who worked to repatriate many stranded seafarers during the pandemic, warned about the implications. The longer shipping companies relied on 'forced labour,' the greater the risk that the hard-won Maritime Labour Convention would be permanently eroded.[7] The crisis could have been resolved quite easily, by designating seafarers as key workers and allowing them to travel internationally, to and from work. It would have cost shipping companies more money, as they are liable for crew travel costs between home and ship, which had increased due to disruptions in air travel.

Tensions rose between workers and their bosses. Labour unions and their global umbrella body, the ITF, prioritised crew welfare and urged repatriation and crew changes as per contract terms. Shipowners, on the other hand, tended to think solely of their bottom lines. Their priority was to deliver cargo on schedule – lest they risk paying hefty penalties to cargo owners for tardiness.[8]

When a ship is delayed, the party responsible for the delay is commonly liable for costs and penalties. This put additional pressure on Cornelius. He regularly asked Captain for 'realistic' or 'accurate' ETAs, which he was unable to provide. He could not, after all, predict the wind, but any costs for missed time slots and delays would be on Cornelius. Given the high uncertainty and additional costs caused by the pandemic, this was not something he could afford.

In August 2020, Kitack Lim, the Secretary-General of the International Maritime Organization (IMO), insisted that the estimated three hundred thousand seafarers then stranded must be repatriated.

'A humanitarian crisis is taking place at sea, and urgent action is needed to protect seafarers' health and ensure the safety of shipping,' Lim said in a statement. 'Overly fatigued and mentally exhausted seafarers are being asked to continue operating vessels, increasing the risk of shipping casualties.'[9]

All the while, most shipping companies were doing well. Thanks to market consolidation, mostly through inter-corporate alliances, shipping rates remained high in 2020.[10] The year after, rates ballooned when container shortages and sky-high demand left cargo owners scrambling for transport options.[11]

As we took on cargo in Veracruz, many seafarers were stranded on board. Some had not left their ships for over a year, causing significant physical and psychological hardship. Gnawing uncertainty enveloped them. When and where would they be able to step off? Would they be able to get home? Would they be able to return to work once they had left? And what would 'home' be like?

On board, offline

'Months at sea with no internet, sailing ship heads back to a "different world"' read the headline of *The Age* reported on my conversation with Melbourne's newspaper on May 1, 2020.[12] It was not the headline I would have chosen. The lack of internet connection was not something I missed at all. If anything, digital disconnection is a boon of being at sea. So thinks my colleague Robert Hassan, at least. He took to the sea to explore what happens when people escape digital distraction and the tyrannical presence of mechanical time.[13]

Cornelius once told me that some people ask if there is Wi-Fi on board, only to lose interest when they hear there isn't. This might explain why no one aboard the *Avontuur* complained about the absence of connection on board: we knew what we were getting into. Most of us, like Hassan, were all too happy to turn off our devices for a while. Throughout the voyage, Cornelius remained in close contact with Captain through the pricey Inmarsat-C connection. Notorious hiccups plagued the system, which is why every email they exchanged was numbered so they could tell if one went missing from the sequence.

For us lowly shipmates, being in port allowed us to reconnect with people beyond our 'contact bubble' of fifteen. It was, indeed, by going online that we first heard of terms like 'social distancing' and 'contact bubble.'

When at sea, we were cut off from the world. *The Age* got that much right. But rather than causing concern or distress, our digital disconnect made us very 'present' aboard.

Connecting our devices meant disconnecting from each other. Captain was glued to his computer and phone to complete the paperwork for cargo operations while speaking to Cornelius at length about voyage plans.

We found it hard to imagine what pandemic life ashore was like. Our landlubber friends and families were fascinated by life in *Avontuur*'s strange form of lockdown. *De Gentenaar*, my hometown's local newspaper, was keen to share the story. Captain, however, seemed suspicious of media engagements. His aversion to media, and especially social media, was obvious.

Few professional seafarers share my desire to disconnect.[14] On World Maritime Day in 2015, participants told the IMO that 'internet on ships [is] key to recruiting and retaining seafarers.'[15] Nihar Herwadkar, a master mariner at Mærsk, wrote an article weighing the advantages and disadvantages of being connected at sea.[16] He argues that access to shipborne connectivity may attract people to work at sea, as it allows them to remain in contact with family and friends. However, Herwadkar concludes, it may not actually be the best thing for life at sea. Social life aboard and rest time may suffer when it's possible to remain in touch with loved ones in real time.

Cornelius, however, fears that when unlimited internet access becomes a requirement under the Maritime Labour Convention – for it's only a matter of time – he will have to close up shop, as satellite connections are prohibitively expensive at sea, particularly when a shipping company does not have a large fleet, which allows it to negotiate deals with providers. This is, most likely, one of the things that will change with the full deployment of Space-X's Starlink satellite service.

In the evening, beer, couriered to the ship by an obscure online shop that operates illicitly within the port compound, would surface from the fridge. Captain turned a blind eye. On the first of these evenings of celebratory drunkenness, Benji admitted to stealing the *Prinzen Rolle* – the disappearing chocolate biscuits that had upset Cook. Maybe they could now get along again?

Ahead of our departure, Captain announced the new watches. He assigned me to the White Watch with the First Officer, Mia, and Joni. The Red Watch now consisted of Captain, Martin, Jennifer, and Pinkie. The Green Watch, led by the Second Officer, contained Athena, Peggy, Benji, and Little Captain. This new arrangement was far from ideal. Personality clashes and conflicts were inevitable. But organising watches when everyone knows each other a little too well is difficult. To assuage the blow, Captain immediately stressed that we'd have another watch change in Horta.

Captain also gave us a task. To familiarise ourselves with a specific ship-handling manoeuvre: we should write down the associated steps in full and discuss it with him or Bo'Sun. At some point later in the voyage, we would put it into practice by giving the commands during a real manoeuvre.

On May 4, my seventieth day aboard, the repaired schooner sail returned. After Bo'sun and Captain inspected it, they immediately sent it back. It would not do.

Meanwhile, I spent countless hours mending the staysail. When patching the holes, I had to be careful not to rip the threadbare sailcloth. Rather than simply sticking patches on, we used Sicaflex to attach the patches before sewing it together with a herringbone stitch.

After lunch, Captain announced that we would call at Helgoland to get the *Avontuur* shipshape before heading into Hamburg. The next day, Captain implored us not to share anything on social media that contravened the ship's strict 'no alcohol' policy. We were, after all, on a working sailing ship that was there to prove a point: sail cargo is feasible. The *Avontuur* is not a cruise ship.

For now, we were getting sick and tired of the dirty port in Veracruz. Cook even came down with tonsillitis, which left her isolated in her bunk. At least we'd been able to keep the coronavirus at bay. Shortly before our departure, we took down the main topmast to improve stability. We had no use for the wooden extension to the steel main mast. Setting the main topsail caused more drift sideways rather than gain speed.

Our imminent departure from Veracruz was a relief. Finally, we'd be returning, rather than sailing further and further away. There was a clear sense of satisfaction that we would make it back to Germany together. The impact on the social fabric aboard was palpable. We'd all slipped into slightly more reclusive versions of ourselves. Our relationship to authority had also shifted. With borders closed and the voyage extending from what most of us signed up for, we no longer submitted to the benevolent dictatorship that is a working ship of our own accord. Few of us had been prepared, mentally or practically, for a voyage this long. Peggy had, but most of us hadn't. We had signed on for shorter stints. And now we were at the whims of the wind and the Captain until we reached Hamburg. I found this difficult to accept. So did most others.

Throughout the return voyage, I protected my sanity by watching movies in my bunk – far less taxing than reading philosophers like Bruno Latour contemplating the apocalypse. I was offline, but no longer fully tuned-in to life aboard. Books and movies circulated more than ever before. I was not the only one relishing being-alone-together. I had soon finished all other books in my sea chest, as well as most of the ones I'd found aboard.

At the same time, I felt the need to weigh in and be part of the bigger picture that is the collective future of mankind. We humans have now taken our planet to a point of return. After decades of campaigning for climate action, we are at a point where we must begin our return journey to a level of emissions, resource use, and pollution that is within planetary limits – even when knowing full

Watching Mathieu Kassovitz' *La Haine* in my 'bunk cinema,' with my laptop suspended from the mattress base of Pinkie's bunk above, July 10, 2020

well that we're in a climate crisis already. Climate action now means urgent damage control.

The challenge with radical climate action, I keenly felt, was that we'd have to accept that some choices would no longer be ours to make. In order to get out of this together, we – that is, rich Westerners like myself – may need to sacrifice cherished privileges that we now think of as our rights.

The COVID-19 pandemic has shown that limiting people's mobility is possible, albeit unpopular. My own travel habits have, I readily admit, long been unsustainable. Over many years, I must have clocked more than two hundred thousand kilometres in the sky. In theory, giving up these conveniences may sound easy. Choosing to do so is possible, but being forced to give them up is a lot harder. Realising that you no longer feel able to opt out of such lifestyle choices is the hardest of all.

The wind is free, but fickle

At 17:00 on Tuesday May 5, we set sail again. Ten days after dropping anchor just outside of Veracruz and after nine days in a dirty, smelly, noisy port, we finally started our return journey.

Once past Isla Pajaros, just off the Mexican coast, we ran the fire pump to wash the deck, scrubbing layers of soot and dust caked onto the ship's steel plates. As I was now on the White Watch, on deck between four and eight, this job was ours. Once the deck was cleaned, I could not resist using the powerful hose for my first proper wash in a week. Joni, now my watch mate, proved yet again to be a stickler for rules: no matter how little work there is to do, one should never wash or launder while on watch.

As we worked in new constellations after the watch reshuffle, our understanding of shipboard life changed. Peggy had not previously realised the sheer amount of work Benji did every day. Though no one seemed shocked about my natural tendency to inertia.

'Low wind. Very humid again,' I wrote in my notes. 'Let's get out of the Gulf of Mexico soon.'

After a week in port, I suffered a bout of *mal de mer* as we set sail. Not really sick, but not quite right either. Bo'sun was at the helm. We all knew he liked spending the first day out of port helming, to get his sea legs again. There was no hiding how we felt. Our feelings and struggles were out in the open. I missed Jen and Benji.

During our first morning watch, rough weather surrounded us. Lightning became more regular, closer, more intense. The wind picked up; the temperature dropped. Rather than sailing northeast towards the Straits of Florida, we were headed due south, back towards the Mexican coast. We weren't just going in the wrong direction; we were sailing directly into the rough weather.

For once, I had a good idea. 'Why not gybe around?' I asked.

This would change our course, so we'd start sailing away from the rough weather, rather than straight into it. Incidentally, it would also change course to where we wanted to head: northeast. My lucky streak soon ended.

The gybe was, as Benji called it, a 'major fuck-up.' He'd started using that expression in defiance of Captain's conviction that 'there are no problems on board, only solutions.' We disagreed. But as Captain insisted we refrain from using the word 'problem,' we came up with synonyms, which he liked even less.

The wind picked up during the manoeuvre and left our short-handed watch scrambling. We were close to an actual major fuck-up. I could not quite get used to the fact that we were not on the *Windspeed*, the thirty-six-foot yacht on which I regularly crewed in Melbourne. Manoeuvres are infinitely more dangerous on a ship like the *Avontuur*. The sheer power during a tack or gybe is incomparable to that of a small yacht.

The movement caused by the gybe woke up Captain, who was now on deck, ensuring our near miss would remain exactly that. The slightest bit more wind and the manoeuvre could have damaged the rigging. I got an earful from Captain for having left the helm for six full seconds, to grab my jacket to protect myself from the cold and rain on the exposed aft deck. 'You are,' he stressed, 'the last safety if anyone should fall overboard during the manoeuvre.' No matter how wet or cold it gets, during such moments, crew safety always trumps comfort. Captain also reprimanded the First Officer for not having called him and Bo'sun on deck sooner. By that point, First Officer was used to not being able to ever do anything right for Captain. He retreated into silence, but it was clear he'd rather be somewhere else; anywhere else.

During our afternoon watch, we gybed again. Now we ended up heading east, towards the Yucatan peninsula. This marked the start of weeks and weeks of making doughnuts through the Gulf of Mexico.

'We're going slowly,' Benji said, 'but in the wrong direction.'

On Sunday May 9, at 16:00 sharp, all hands gathered on the aft deck for the Captain's Reception. It was the first time all fifteen of us were together since leaving Veracruz five days prior. The sip of rum served was 'good stuff,' carried on the *Avontuur* from the Caribbean the year prior. Even so, German tradition dictates the first sip go to Rasmus, the patron of seafarers, to encourage good weather. For every second Captain let the rum flow into the ocean, we hoped to get one knot of sailing speed. Our failure to swiftly sail out of the Gulf of Mexico proved this technique far from effective. Engines make better propulsion technology than superstition – though superstition is never far when at sea.

'Allah is great,' Allan Villiers – the Melbourne-born sailor who documented the demise of commercial sailing ships – said to Nakhoda Nejdi when they met again in Kuwait in 1967, after sailing together in 1939.

'Allah is great,' Nejdi responded, after which he continued the conversation, seemingly more with himself than with Alan Villiers. 'Sometimes I wish that I could use his winds again. For it was a good life that my sons can never know – no Kuwait son shall know.'

Villiers agreed, 'We cannot bring those ways back again.'[17]

The men reminisced about the last days of sail-powered trade in the Indian Ocean. As they met again, two decades after sailing together on the brink of the Second World War, oil exploitation had transformed the Arabian Gulf. By 1967, 'The Gulf' was a metonym for the oil industry, a region with very few sailing dhows left. Dubai's Jebel Ali is now the biggest port of the region, while de-masted dhows serve as tourist attractions in the inner-city Dubai Creek.[18] The Second World War was, indeed, the war that created the technological innovations and investments in infrastructure that saw the boom in oil-fuelled ships.

After the war, they became virtually the only means of maritime cargo transport. The rise of the oil-burning ship and the associated shifts in logistics depended on military innovations, both technologically and organisationally.[19]

How surprised would Alan Villiers and Nakhoda Nejdi be to see the revival of sailing cargo vessels? In defiance of economic logic and regulatory constraints, long-dead sailing cargo vessels are now being offered a new life. While the *Avontuur* and sister vessels prove that it is *possible*, I often pondered how this return of sail power reflected the free, but fickle, nature of winds at sea. Our bad luck with the wind on *Avontuur* is precisely why the shipping industry is so reluctant to embrace the wind. Our troubles were not with the wind per se, but with a particular vessel in the particular environment. No sailor in their right mind would suggest sailing out of the Gulf of Mexico in a flat-bottomed bathtub like the *Avontuur*.

As shipping makes globalised just-in-time supply chains possible, there is little leeway for delays. Those minor delays pale against the rising uncertainties about fuel supply and cost. Jean Zanuttini, the CEO of Neoline, a Nantes-based French company that is planning to build a modern sailing cargo ship, thinks the uncertainty of the wind is negligible. In an interview with Rhys Berry for *Bunkerspot*, a shipping industry newspaper, he explains why:

> We are much more able to predict the prices because we are much less impacted by any kind of fuel move and so we completely avoid bunker adjustment factors and we can take multi-year engagements without any terms of valuation on this aspect. And that's one of our selling points; to be able to give this kind of insurance.[20]

This does not mean that wind propulsion is without its problems, as Gavin Allwright, the Secretary-General of the International Windship Association told me when we first met in London. 'One problem with wind propulsion is you can't commodify it. So, you sell a wind propulsion system, that's it. If you sell a hydrogen system, for the next fifty years you're selling hydrogen to that system. That's where you make your money, not from the hydrogen system, from the delivery of each unit of hydrogen.'

He counters this argument by pointing out that 'wind propulsion solutions lock in a percentage of fuel costs at zero, creating certainty where there is none. We can't foresee future fuel prices, policy frameworks, carbon pricing or rationing.'[21]

However, shipping companies use contracts that put fuel costs onto the ship's operator or cargo owner, rather than the shipowner. This disincentivises owners to invest in fuel-saving technology[22] – especially if fuel remains cheap, in the absence of a carbon price. The situation is similar to expecting home owners to invest in energy-saving renovations that will save their tenants money.

Beyond the unpredictability of fuel costs, Cornelius resolutely opts for sail as the primary means of propulsion:

> My intention was always to use a 100 percent clean propulsion system, as a main propulsion for the ship. That means sails are the main propulsion, and there is an auxiliary drive like an electric engine that is driven by a fuel cell or something. But that's auxiliary and not the other way around.

Using wind as the primary means of propulsion requires skill. A skill that most seafarers no longer learn. Our First Officer was a skilled and experienced seafarer. But he had been out of work, by choice, for the past four years. This was his first time back at sea. There he was, stuck with a bunch of hippies on a sailing ship. It wasn't just a matter of cultural difference. Sailing, unlike cinema and literature, was something totally foreign to him as he joined the *Avontuur* in Tenerife. This lack of sail handling skills is common among professional seafarers today.

The Dutch *Enkhuizer Zeevaartschool*, or *Enkhuizen Nautical College*, Europe's leading school for training seafarers to operate under sail, has trained the crews of 'sail training' vessels that operate adventure, expedition, and education vessels since 1978. In collaboration with the EU-funded Wind Assisted Ship Propulsion

Initiative, the school started offering a course in wind-assisted propulsion in 2021.[23] While aimed at seafarers operating Flettner Rotors, rigid sails, and kites on conventional cargo ships, it could help people like our First Officer, who knew how to operate an enormous container ship, but had no experience trimming sails.

'My television was my best friend,' said the First Officer in reflection on his childhood. After joining his watch, we started to discuss art, culture, and travel. He was clearly a cultural omnivore. But underneath his well-travelled life and general cultural appetite hid a melancholy person who seemed to have lived richly, yet enjoyed little. He could describe spending years of his life cycling through Europe, without mentioning almost anything he actually liked about it. Perhaps he was the absolute master of understatement, though he did seem to have lost his lust for life. He never wanted to work on a sailing ship again. Perhaps, he'd even give up the merchant marine altogether.

At face value, the First Officer was in charge of the White Watch, but given his lack of experience, deckhand Joni would whisper suggestions and ideas. No one considered his suggestions optional, but we upheld ship's hierarchy as a never-ending theatre in which Joni pretended that he was the First Officer's subordinate. Even so, we were often at pains to convince the First Officer to call Captain on deck, which he had to do before initiating any manoeuvre.

This was not the only issue we faced. Well over two months into the voyage, our manoeuvres remained slow, sloppy, and riddled with errors. Some of this was for lack of practice, as we'd gone days without any manoeuvres on the westbound Atlantic crossing, but some was down to fatigue and tension aboard.

Shipping more, polluting less?

'Global capitalism is a seaborne phenomenon,' Liam Campling and Alejandro Colás open their book, *Capitalism and the Sea*.[24]

Production and consumption are oceans apart, so business can profit from income differences between workers and consumers.[25]

If the global economy is to keep on growing, so must the shipping industry. This is the often-heard excuse of shipping industry executives: 'We merely offer a service to the economy; we do not grow for our own sake, but to support the prosperity of economies and people.' This is my loose interpretation of the myth that pervades the shipping industry, though it's effectively what the former IMO Secretary-General told the participants of the Future-Ready Shipping Conference in Singapore, mere months ahead of the Paris Agreement.[26]

This much may be true. But repeating this excuse stops them from discussing the scale of their industry. The unquestioned mantra of perpetual economic growth distracts us from the bigger question: can the global economy keep growing on a finite planet?

In *Societies Beyond Oil*, the sociologist John Urry discusses the implications of running down our resources and the need to phase out fossil fuels to mitigate further anthropogenic climate change. He outlines four potential scenarios for a future in which oil will be scarcer and pricier due to the rapid depletion of remaining reserves.[27]

The first scenario, a magic bullet future, relies upon quickly and almost entirely reducing fossil fuel reliance. In this scenario, alternative energy sources cover current and future needs while allowing for the expansion of transport. New technologies such as gigantic zero-emission vessels fall into this category, provided they do not rely on upstream emissions.[28] This vision of the future aligns with 'ecomodernist' views on environmental action and regulation. It necessitates no rethinking of consumption or the extent to which mobility of people and goods is possible. The political economy of shipping in this model remains essentially unchanged.[29]

It is in this context that headlines promising that 'more than a quarter of all tonnage under construction will use alternative fuels' make sense. Never mind that most of the promised alternative fuels are still fossil fuels.[30] Innovation is, after all, the solution. Never mind that when we take the slippage of liquefied natural gas (LNG) into account, this fossil fuel is as damaging to the environment as oil is. Never mind that methane, which makes up most of LNG, is a far more potent greenhouse gas than CO_2. Even tiny amounts of 'slippage,' the leaking of gas during transport or storage, results in a significant greenhouse gas effect. The NGO Transport & Environment conducted a spot check of LNG-powered vessels in the port of Rotterdam, which led it to conclude that rather than helping fight climate change, methane slippage might actually make things worse.[31] This is precisely why the Norwegian Minister of Climate and Environment Espen Barth Eide clearly stated that the time LNG could serve as a transition fuel is now over.[32]

Meanwhile, 'the industry's preference for regulation based on available technologies could be an issue,' argue Aldo Chircop and Desai Shan, 'because GHG regulation should be proactive and foster an environment for the development of new technologies, rather than be reactive and responsive to technological availability.'[33]

To make matters worse, pending the full deployment of zero-emission technologies and fuels, the Jevons Paradox perpetually lurks. In 1865, the economist William Stanley Jevons wrote that innovation could lower the resources and energy needed per unit produced. Since the onset of the industrial revolution, such innovations have often led to lower prices per unit – meaning that shipping costs could decrease as a result of increased fuel efficiencies, which drives up demand. It can cause a rebound effect, whereby relative reductions in resource use (emissions) may drive up aggregate use (emissions).[34] Today, the 'cost' of shipping should include its environmental impact, as cargo owners start expecting lower carbon emissions. This means that the Jevons Paradox might apply to the environmental cost of shipping, even if the monetary cost increases. This is a possible scenario for the shipping industry, which could deplete the remaining carbon budget ahead of full decarbonisation – despite all expected efficiency gains.

Furthermore, even if full decarbonisation happens, shipping faces further challenges: decoupling economic growth from its material footprint has, thus far, remained even more challenging than decoupling from carbon emissions.[35] Iron ore, for example, is used to build ships but pollutes the environment in many ways beyond carbon emissions.[36]

John Urry's second scenario for a world beyond oil – digital lives – relies on a near-complete reduction of transport needs by shifting to remote working, online experiences, and decentralised 3D manufacturing. For this scenario to succeed, transport and travel have to be reduced almost entirely. If overall reliance on fossil fuels decreases, this would immediately diminish demand for cargo transport, as more than a third of maritime cargo by mass consists of fossil fuels, including coal, oil, and gas.[37]

During the COVID-19 pandemic, reduced personal mobility became a reality for many. However, it is unclear if this is a sustainable long-term option. While COVID-19 has significantly reduced demand for passenger transport, freight transport remained stable in 2020 and increased significantly in 2021.[38] Even if demand for cargo transport diminishes in the wake of a sustained economic downturn, it is unlikely that this will result in sustained drops in emissions. Indeed, both the IMO and industry observers report that lower cargo volumes would slow the deployment of new vessels that are more energy-efficient while lower oil prices make fuel-saving investments less profitable.[39] Furthermore, a systematic review of studies on 'telework' suggests that the emissions reductions are, on the whole, modest and sometimes negative, not least because 'working from home' relies on 'energy-intensive forms of digital technologies' and 'rare earth metals and minerals,' which all require shipping.[40] Being stuck at home led to increased online shopping, which in turn increased demand for shipping. Moreover, the unequal dispersion of food production and populations raises questions of how to feed the urban majority; we can't, after all, live digital lives on digital food.

Urry argues that, as the first two options are relatively unlikely, business-as-usual would lead to the undesirable third option, resource fights, which is by far the most pessimistic. He forecasts worldwide geopolitical struggle over rapidly diminishing oil resources. The dregs will be increasingly difficult and expensive to obtain, leading to a real-life planetary Mad Max. Urry stresses that because geopolitical tensions over resources are already widely spread, 'most economies will be unable to cope with future increases in oil prices.'[41] Worryingly, he argues it is unlikely that countries will be willing to pursue common interests by pooling scarce resources. This lack of cooperation will further exacerbate geopolitical tensions, as Russia's invasion of Ukraine on February 24, 2022 has so painfully illustrated.

A fourth option involves shifting to a low-carbon society in which we will 'live smaller, live closer, and drive less.'[42] This view of society echoes the message of Tim Jackson's post-growth manifesto, *Prosperity Without Growth*, in which he argues that no economic growth does not necessarily equate to less prosperity.[43] This scenario goes against the pursuit of ever-increasing growth, consumption, and material wealth. It may not, however, be the desired option for many. Even its proponents, like Urry, admit that if this is the future we want, 'the key questions are how to get to such a powered-down future and how to get there fast enough.'[44]

Decarbonising societies clearly isn't a purely technical question. For shipping, too, the question of which future to pursue is a political one. The debate keeps returning to the same question: can the global economy keep growing on a finite planet?

Some argue it can, by claiming that technological innovation will help us 'decouple' economic growth from resource use by doubling down on economic growth driven by innovation.[45] Such innovations should allow emissions to decrease while economic activity increases. Others argue the planet cannot cope with infinite growth.[46] They suggest we – rich Westerners, that is – need to 'degrow' the economy. Such ideas align with the arguments of environmental activists like Naomi Klein and Bill McKibben.[47] Given the urgency of decarbonising the industry, we need to ask whether pursuing wind propulsion can help the shipping industry reach net-zero emissions as soon as possible.

Changing the means of propulsion will not change unsustainable levels of consumption. If anything, cleaner shipping will, following the Jevons Paradox, lead to greater demand for shipping. Greater demand, in turn, will fuel further production and consumption. For all their limitations, sail cargo initiatives aim to both decarbonise shipping and encourage reduced consumption – but will they succeed? Much like our voyage, humanity is faced with a simple question: what kind of future do we want?

Captain commented that 'Cornelius is looking at us living his dream.' But that dream was turning into a slog. Day in, day out, we were doing the same. Sleep, eat, stand watch, repeat.

We started being ourselves a lot more. In our honesty and openness, we no longer pretended to be on board primarily to ship cargo. We were there for a range of diverse reasons. I was fleeing my comfortable but somewhat predictable university job. Others wanted to change direction in their lives. Much like Urry's scenarios, we looked at the present and the future differently, causing tensions in daily interactions and in terms of the 'bigger picture' of a world facing a climate emergency, of which we seemed to be a microscopic parallel universe. We faced the same questions of energy, consumption, waste, feasibility, and sense of reality. Our voyage put our principles into practice in a way that life ashore never could.

Changing weather

On May 9, we were sailing at seven to eight knots during our morning watch. When Captain came up, just after sunrise, he immediately ordered us to reef the schooner sail and douse the main. In the dark of the night, we'd already doused the outer and flying jibs. Before anyone had breakfast, we sailed straight into a thunderstorm. Force 9, gusting 10.

'Sublime,' I noted in my diary afterwards. 'Lots of rain, a big swell, and lightning.'

Captain ordered us inside. Locked in the galley, a Faraday's Box, we were safe from any lightning. But the crowded space offered no respite from the heat and humidity. The *Avontuur* has no below-decks connection between the fo'c's'le and the aft cabins. Rough weather would thus trap you on one side.

At times like that, stories of voyages past would surface. Bo'sun and Joni were *Avontuur* veterans and warned us of the notorious North Atlantic we were soon to enter. So far, we'd seen more calms than rough patches. On this day too, the storm abated by 10:00, three hours after its onset. Come afternoon, the sun was out again, and we were sailing due east at a slow but steady pace. If anything, we looked forward to some stronger winds and rougher seas. It was time for some adventure.

As we tried to leave the Gulf of Mexico, the weather was a mix of calm winds and squalls. Most often, the wind did not really come from the right direction. Rather than smoothly making our way out of the Gulf, we spent nearly every watch setting and dousing sails, reefing in and reefing out.

The prevailing winds in the Gulf during the month of May are easterlies. With a vessel that can't sail close to the wind very well, that's a bit of a problem, despite Captain's aversion to the word.

With a modern yacht, we'd have been able to sail close to the wind far better than with our flat-bottomed schooner. Her rig did not allow us to sail closely upwind. The flat bottom of the hull made it worse, as she'd drift sideways through the water when doing so.

Bo'sun told me there had been plans to weld a lower keel to the vessel to make her track better through the water and record a better course-over-ground. Much like the planned prop-generator, which would turn the kinetic energy of the propeller's spinning as she sails into electricity, a lower keel remained a mere plan. A differently shaped hull would have made getting out of the Gulf of Mexico a whole lot easier. Then again, so would an engine.

Between the constant sail and course adjustments, we were putting in a lot of work on deck. As these manoeuvres required both Captain and Bo'sun on deck, they looked utterly exhausted within a week. As Captain trusted neither First nor Second Officer to manoeuvre the ship on their own, they had little choice but to forego sleep.

The next day, May 10, low winds made for violently flapping sails. This causes far greater strain on both sailcloth and rigging than having strong wind in the sails, as we had discovered when the schooner sail ripped just before arriving in Veracruz. Shifting winds, caused by a passing squall, led to an accidental tack. After that, we were heading west on a northerly wind. Once again, we were sailing in the wrong direction. Gybing around brought little resolve, as the wind died out mid-manoeuvre.

Later that day, the First Officer asked me what exactly appealed to me in sailing. By this point in the voyage, I wasn't sure anymore. But whatever enticed me about sea and sail, he did not share it. His bunk, at the base of the main mast, made for noisy sleeping. The cables running up to the radar passed through the hollow core of the steel mast and slammed on its sides at every roll or pitch. Unlike the Second Officer, with whom he shared a cabin, he barely slept at night. Napping on deck in the afternoon helped little.

Oftentimes, we'd be far too numerous for a manoeuvre. This would lead to overcrowded 'stations,' making the work harder as we stumbled over each other. All the while, overeager men would often jump stations and overpower those already hauling lines, much to the dismay of those – often women – already looking after that station. Even when not on watch, you'd be pulled into each and every manoeuvre when on deck. I'd rather hide in my bunk with a book than seek repose on deck, for I was getting frustrated with the expectation to play busy at all times.

Some days later, while I was at the helm and Pinkie went forward, I asked her to close the fo'c's'le hatches ahead of an imminent squall. The First Officer, who was standing next to me, went forward with her and closed the hatches. Pinkie may have been short of posture, but she needed help from no one.

More exhausting than physical labour was the emotional labour to ensure the social fabric did not disintegrate. The constant threat of tension and conflict meant that we worked to mend and maintain social ties. As a result, little time was left, as Jennifer mentioned to me, to get into a reflective mood. I felt much the same, as time to think, wander, reflect, and retreat into oneself was increasingly hard to find.

This may sound odd to anyone who's never lived aboard a ship on the high seas. But social life is intense when space is at a premium. There isn't much time to ponder and reflect on the horizon and the stars. The close proximity to others means that anything can happen at any time.

We spent a vast amount of time on repairs and maintenance, simply to keep things in an acceptable state. Nothing is ever really repaired. But we couldn't allow anything to disintegrate further. Through care and maintenance, we can mend the ship, our gear, and our social lives.

'The repair worker,' of both the ship and its social fabric, 'has to treat breakdown as a caution as well as an opportunity,' Richard Sennett reminds us in his book *Together: The Rituals, Pleasures, and Politics of Cooperation*. 'When an object goes wrong, we need to think about what was wrong as well as what was right about it in the first place.' I will later return to his careful reflections on the politics of living and working with others, as they help us to understand both the joys and the difficulties of sharing a world with people who think differently.

There were sublime moments on deck during squalls, the ever-changing weather and winds. The constant shifts in our social life made the ship feel different hour by hour, day by day. We moved slowly, but time passed quickly. We'd been aboard for months that felt like lifetimes.

'A sailor's life,' Henry Dana Jr reminds us, 'is at best but a mixture of a little good with much evil, and a little pleasure with much pain. The beautiful is linked with the revolting, the sublime with the commonplace, and the solemn with the ludicrous.'[48]

It was at this point that I came to understand Dana's longing to cut his voyage short. The excitement and novelty of seaborne life wears off quickly. The difference is that today we're all far more vocal and individualistic. No one would now accept the kind of treatment that Dana faced nearly two centuries ago. Even so, we could no longer pretend to be aboard willingly, of our own volition. We were no longer sailing for fun. We were simply working – most of us without pay – like clockwork, for eight hours a day to keep Cornelius' project afloat.

After about twelve days at sea, since leaving Veracruz, murmurs emerged. The weather information Captain was starting to get was less reliable. We had barely made a dent in the distance we needed to cover to get out of the Gulf of Mexico.

If we had been on a 'normal' cargo ship, making a constant twenty-two knots, we would have been in Hamburg by that point. It would take us, on our sailing ship, far longer than that. But worst of all, we had no way of predicting when we'd arrive.

At lunch, I mentioned to Captain that I had started watching *Chernobyl*, the HBO mini-series that recounts the 1986 explosion of the nuclear reaction near Pripyat, in present-day Ukraine.

'You must be bored,' he responded, 'if you have started watching such things.'

I was not bored at all. I just needed time on my own. And my bunk was the only place where I could find that peace.

Meanwhile, I had to remind myself every day why I was there: I was trying to understand how sail cargo initiatives actually work in practice. That part was easy.

I also hoped to hear others' perspectives. This part was trickier, as we'd settled into our lives and personalities. Into our habits and routines. It was difficult to regain the initial enthusiasm and excitement we experienced on our westbound Atlantic Ocean crossing. I had gone native. As we slowly settled into our own 'new normal' of the return voyage while locked out at sea, I started speaking to my fellow crew about their experiences and thoughts more often. I started with Peggy, who had joined the ship for similar reasons – though hers was a far more activist project than my own.

A good Anthropocene?

Peggy is an environmental activist and a staunch feminist. Throughout the voyage, she was one of the most outspoken advocates of Timbercoast's 'mission zero' cause. She quit her job as a primary school teacher to work in 'environmental protection,' while educating people through her experiences. A year after we returned from our joint *Avontuur* voyage, she embarked on the *Gulden Leeuw* (*Golden Lion*) to engage with high school students who will spend a year on the water with *Class Afloat* – one of several companies organising year-long education programmes at sea.

Peggy is eco-pious – as the scholar of environment and religion, Sarah McFarland Taylor, would call it – an environmental idealist, committed to doing the right thing. Peggy's motto, 'we are the change,' reflects her commitment to social justice, but also her stubborn belief that our small eco-conscious actions are making a difference.

When I asked her what her ideal future looked like, what she's working towards, she responded that 'the better future would be, to me, that everybody is just open-minded and give their hearts.'

After our encounter with the sixteen refugees stranded aboard the *Marie*, she said 'we should have an equal world so this kind of thing would not have to happen.'

'What would that mean in practice?' Athena asked her right away.

'Rights and respect,' she responded, failing to translate her ideals into workable and pragmatic actions. This kind of thought and speech, lacking any 'real world' application, sums up Peggy's approach to climate change, too.

Reading book after book, report after report, the message has been the same for the past few decades: we are very close to irreversible tipping points; but if we act now, we can still avert the worst climate change. How can you say that decade after decade and still be right? Will we be able to avert the worst effects of climate change or are we doomed? It's hard to believe that we stand a chance to reverse the course of history we've embarked upon. And *if* we can technically pull it off, will we make the necessary sacrifices to make that shift? If anything, the COVID-19 pandemic as it unfolded – far away back on land – seemed to indicate that rapid change was possible. Yet the powerful 'merchants of doubt' hampering climate action at every step, are preventing such drastic action for the health of the planet.

In an early commentary on the pandemic, the French philosopher Bruno Latour observed how it did the impossible: showing that we can in fact 'put an economic system on hold everywhere in the world at the same time,' even though we were told, in the context of climate change, that it was 'impossible to slow down or redirect it.'[49] Can we accomplish and sustain a similarly drastic set of actions for climate change?

Even if the shipping industry could, at least theoretically, attain 'zero' emissions and continue growing, that does not mean that pursuing endless growth of shipping, and the economic activity it enables, is sustainable. Meanwhile, the lure of marginal efficiency should not distract us from the real target:

zero emissions. While changes that lower the carbon intensity of ships are necessary in the immediate term, 'simply placing faith in technical and operational measures, such as ratcheting the Energy Efficiency Design Index (EEDI), a mandatory IMO indicator that reflects the carbon intensity of a ship's operation, to higher levels, is likely insufficient given the expected growth in world trade.'[50]

More importantly, it is unclear what the unintended consequences of current innovations may be. While the benefit of hindsight favours our assessment of the past over the future, we have been confronted with significant unintended and unwanted side-effects throughout industrialisation and globalisation. It is difficult to predict what kinds of negative side-effects further innovation will have. This is not a reason to stop innovation, but a reminder that its consequences won't be unequivocally positive.[51] The side-effects of industrialisation, of progress, and of modernity now mean that humans are what defines the current geological era.

'Stop saying the Holocene! We're not in the Holocene anymore,' said Paul Crutzen at a conference in Cuernavaca, Mexico. 'We are living in the Anthropocene.'[52]

This off-the-cuff suggestion resonated so much with scientists and environmentalists that in 2019, the Subcommission on Quaternary Stratigraphy decided to formally adopt the term for the current geological era, which they agreed started around the middle of the twentieth century.[53]

Little disagreement exists over the threat of runaway climate change. Few refute that humans are indeed responsible for this crisis, though there is great disagreement on how to get out of this pickle.

Some proclaim that the Anthropocene shows the power of human domination of nature, which we should by all means continue, expand, and extend in order to combat climate change. These are the 'ecomoderns' who believe the solution is more of the same, but better.[54] To them, the Anthropocene is intrinsically good, as the proof that humans can in fact change the climate – and can do so again.

Others think the Anthropocene shows nothing less than humans' inability to live on planet Earth without destroying it. Our collective pressure is so immense that the only way out is to slow down, consume less, and reduce the environmental pressure on the planet. These are the 'degrowth' proponents, who argue that we should be able to live in prosperity with less.[55] To them, the Anthropocene is intrinsically bad; it is proof that life is out of balance.

Godfrey Reggio, an experimental documentary filmmaker, made exactly that point in his 1982 cinematographic essay *Koyaanisqatsi*. In this feature-length film, the pace of life becomes ever-faster and ever more industrial and urbanised. Accompanied by the accelerating arpeggios so characteristic of Philip Glass' work, the visuals tell a story that was as important then as it is today. The title, in case you were wondering, is a Hopi language expression that means 'life out of balance.'

The problem is that these incompatible views of what is happening propose competing solutions for the future of humans on Earth. The same goes for the future of the shipping industry. Gavin Allwright of the International Windship Association argues we're in a three-part wave of innovation. First, *tweaking* the existing fleet by retrofitting ships that use fossil fuels with wind-propulsion technologies which allow for incremental reductions of carbon intensity. These include Norsepower's Flettner Rotors and eConowind's rigid sails, which can be mounted as a fixed installation or fold out of a forty-foot container latched to deck. Second, a *transitioning* to new vessels that promise to reduce fuel use by up to 90 percent, such as *Oceanbird*, *Canopée*, and *Neoline*. These wind-propelled and wind-assisted vessels are meant to ensure the industry as a whole is 'wind-ready.' While they are promising, none of the 'primary' wind-propelled vessels are in operation yet – and it will likely take until 2023 until the *Canopée* will be ready to sail.

Third, *transforming* the industry, so that primary wind propulsion becomes the standard option for most vessels. This is precisely what Gavin and the IWSA membership are working towards by calling 2021–2030 the decade of wind propulsion.

When I asked him if this is environmentally viable when the shipping industry and the global economy keeps on growing, I pointed out the risk that the Jevons Paradox might undo all the fruits of wind-powered innovation.

'Gavin's Paradox, yes, that is the story of my life,' he quipped. But quickly added in all seriousness that he 'can't be concerned with' sustainable shipping beyond zero-emissions. This is not because he doesn't care about these issues, but because the immediate objective of the International Windship Association is to speed up the deployment of wind-powered propulsion technologies because they exist and because they work. Retrofitting ships with such devices as quickly and as much as possible can help harness the power of the wind that is so abundant at sea.

Every last person I've spoken to is concerned about the future. Every single person is doing *something*. Peggy, for example, clarifies that her 'main reason to be on this ship is to send out the message that we have possibilities.' She wants to spread a message: 'We are not stuck. A lot of people talking to me, say "oh, what can I do?"'

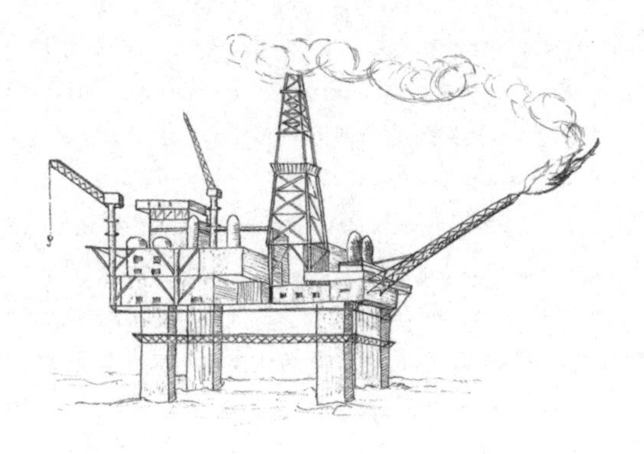

6
The eternal frontier

Many deepwater horizons

I first saw an oil platform in the Gulf of Mexico on May 15. It was the *Rowan Relentless*, an 'ultra-deepwater drillship,' which can bore wells as far down as ten thousand metres, in water as deep as over three thousand metres.[1] It can reach oil as far below the surface as a jet plane flies above it. The vessel is owned by ARO Drilling (a subsidiary of Aramco, a Saudi Arabian oil company, the world's largest) and is operated by ExxonMobil. It uses Dynamic Positioning to remain in place using a combination of sensors and satellite systems, which feed into mathematical models that drive multiple propellers and thrusters to remain exactly in position. It is a very precise, but highly fuel-intensive manner to keep the ship in place when jacking up a barge or anchoring is not possible due to the great depths. As the *Rowan Relentless* cannot move without compromising its operations, it asks ships in the vicinity to maintain a distance of at least two nautical miles. Rather than calling other vessels over VHF radio, they simply include the message in their AIS information.

The *Rowan Relentless* was the first of many such platforms we saw. The same day, I spotted *Mad Dog*, *TLP Stampede*, *Green Canyon*, *Spar Tahiti*, *TLP Shenzi*, *TLP Seastar*, and *Anadarko*. The latter is named after a Texan oil company that owned a minority share in the Macondo Prospect, the site of the Deepwater Horizon disaster. The platforms I saw were just a fraction of the more than sixteen hundred that the US Home Bureau of Safety and Environmental Enforcement lists as operating in the Gulf of Mexico.[2] Even that enormous number pales when considering the six thousand 'structures' that the National Oceanic and Atmospheric Administration says have been installed since 1942.[3] Of that number, three thousand two hundred remain active, and another two hundred remain present, albeit inactive.

The presence of so many oil platforms illustrates the sheer scale of oil extraction. These platforms are normally only visible to most people when things go horribly wrong, such as the explosion of the Deepwater Horizon. But shipping, too, is increasingly invisible, as the vastness of the oceans is too large to grasp.

Rachel Carson's 1962 best-seller *Silent Spring* was a watershed moment in thinking about the environment. Her book turned concern for the environment into a mainstream issue for middle-class citizens in the USA and elsewhere. Carson's argument was based on scientific knowledge but communicated through an imagined 'silent spring' that threatened the environment if agriculture continued to use the pesticide DDT. Whether it is nuclear waste or 'forever chemicals' in the soil and oceans, the risk is real but invisible. Today, we remain largely 'sea blind' to what happens beyond our shores. It's hard to imagine the sheer size of the ocean when standing on a beach.

As oil reserves on land dwindled, oil extraction became riskier as it moved further offshore. To keep high-carbon societies supplied with oil, petrochemical companies have drastically increased risks of major pollution. The Gulf of Mexico will always remind me of the explosion of the Deepwater Horizon, BP's Macondo project drilling platform operated by Transocean Ltd. The explosion on April 20, 2010 happened about ten years before we crossed that very area aboard the *Avontuur*. The disaster spewed 780 million litres of crude oil into the sea.[4] Despite the use of millions of litres of chemical dispersant meant to keep the oil below the surface, the spill 'resulted in an oil slick ultimately covering more than 112,000 km^2 on the ocean's surface' and led to 'oiling along 2100 km of shoreline.'[5]

As we sailed through the area, I washed myself, my clothes, and our dishes in that water.

Peggy spotted 'dumping grounds' on the chart during night watch and was surprised that disposal of all kinds was permitted at sea. Mia was shocked and seemed baffled by the idea. She even suggested – as we had no idea what was dumped below – that it would, perhaps, be 'better to skip the bucket showers for now.' My reply that the dumping grounds were a full three kilometres below us made no difference to her.

Their concerns struck me as somewhat naive: pollution is part and parcel of economic life at sea. But perhaps I should have heeded their shock and concern. Perhaps I had fallen for the carefully crafted spin of polluting companies.

MTI Network is a PR agency for the shipping, energy, and offshore industries. It claims that '99.9% of crude oil' was 'delivered safely last year.'[6] In 2019, the total amount shipped was 2 billion tonnes, or 2 trillion litres. If only 0.1 percent, or 1/1000th, of that amount did not arrive safely, it adds up to an astonishing 2 million tonnes, or 2 billion litres. That's not exactly a number to be proud of.

Meanwhile, on board the *Avontuur*, the rapidly diminishing planetary carbon budget was not our primary concern. On May 16, my watch mate Joni was grumpy. Someone had baked bread, which was not part of the plan. Cook's assessment of the remaining flour and gas supplies signalled low stocks, even though we were only two weeks out of Veracruz. There are no shops at sea. So, we'd simply have to bake – and eat – less bread, lest we risked running out of gas and flour before reaching Horta.

If only reducing fossil fuel consumption was as easy as baking less bread.

The climate targets we now have focus on temperature ranges. Pacific Islanders insist on keeping warming within 1.5 degrees above pre-industrial levels, as this means they might be able to save their countries from inundation. Others stick to the less ambitious 2-degree target. But even that is not within reach, as the promises made in Glasgow at COP26 put us on a 2.4 degrees trajectory.[7] The likely level of warming is far higher, as such promises have rarely been kept in the past.

One big planetary challenge is knowing just how big our remaining carbon budget is. Intergovernmental Panel on Climate Change (IPCC) estimates range from 300 to 2300 gigatonnes of carbon dioxide. Sticking to the 300 gigatonne budget gives us an 83 percent chance of keeping temperatures below 1.5 degrees. Sticking to the 2300 gigatonne budget gives us a 17 percent chance we will see temperatures increase by no more than 2 degrees.[8] The caveat? These budgets were valid at the start of 2020. By the time you read this, we'll have an even smaller carbon budget left.

Whether it's bread supplies or the climate, we have to predict the future based on incomplete information. In our case, aboard the *Avontuur*, the most pressing question was time, the time it would take to reach the next port.

When I told Joni I was unaware of the decision to curb bread-baking, he said it was none of my business. I disagreed. As we were all in the same boat, we had a right to know our collective plans.

'How can we be a community,' I asked, 'if we don't even know about the decisions that directly impact us?'

A few hours later, Joni told me he was in a bad mood that day. But he was only trying to help Cook. Having worked as cook aboard the *Avontuur* on an earlier voyage, he understood how challenging it is to put food on the table for fifteen people, three times a day.

'If governments are serious about the climate crisis,' Fatih Birol, the executive director of the International Energy Agency, told *The Guardian*, 'there can be no new investments in oil, gas and coal, from now – from this year.'[9]

That was in May 2021.

In the same week, the Australian government announced twenty-one new offshore exploration areas in the Great Australian Bight, along the country's southern coastline.[10] The so-called 'net-zero' plan the Prime Minister pulled out of his sleeve days before the COP26 in Glasgow doubled down on that. Australia remains committed to fossil fuels until 2050 and beyond.[11] Currently, there are more than one hundred new fossil fuel projects planned in Australia alone.[12] If they all proceed, this would add up to 1331 million tonnes of carbon in additional emissions per year.[13]

On May 16, around 05:00, we were seven nautical miles from the *Rowan Resolute*, another enormous deep-sea drilling vessel. She was surrounded by the *Maggie A* and *Ella G*, two support vessels, and the *C-Endurance*, a supply ship – a team of floating helpers that are instrumental in wrecking the planet. Without them, it would be impossible to pump up submarine oil.

I asked the First Officer what our CPA was. That's short for Closest Point of Approach. For those unfamiliar with shipping navigation, this is a rather crucial numerical value that spells out just how close you will come to another vessel when maintaining course and speed.

'We're going straight at her,' he laconically responded.

We were on a collision course with a drilling ship that cannot move. In the dark, the *Rowan Resolute* looked like a floating casino. Its bright lights might have seemed rather inviting, if Las Vegas is your thing. But, inviting or not, it's never a good idea to get too close at sea.

Still, we couldn't change course much either. We could not give her more leeway, for we were already sailing as close to the wind as possible. So, we passed the ship on her lee side.

Over the radio, Captain told them we'd pass their stern. But as she was moving to and fro to remain in place, it was her stern that faced windward and moved slightly in that direction.

Rather than calling them again to correct our mistaken observation, Captain turned on the engine and rounded the vessel to windward. But even with the engine compensating for our leeward drift, I could not keep our course high enough. Joni took over as Captain said that 'trying is really not good enough' in this case.

We spent an hour motor-sailing for no other reason than to avoid a drilling platform. Sailing would be a whole lot safer and easier if no such fossil fuel pumps littered our seas.

Freedom of the seas

In 1609, the Dutch legal scholar Grotius published *Mare Liberum*, commonly translated as *Freedom of the Seas*. This treatise proclaimed the right, particularly of the Dutch, to roam the seas in search of profit. Using the seas for transport would not harm anyone, or so his thinking went. Never mind that the Dutch East India Company certainly did harm people, on land and at sea.

Our encounter with many of the thousands of drilling platforms that litter the seas told us a different story, and the imminent deep-sea exploration for mining tells an even more frightening one. The sea, despite its immense beauty, is no longer the eternally sublime space that romantics once imagined it to be.

On June 1, the ocean looked more beautiful than it had at any point in the past few months. There was a good swell, thanks to a sustained wind for several days. The sunlight sometimes gave the waves' crests a turquoise translucence just below the whitecaps, which topped nigh all of them. The sea state was pretty rough, but it felt utterly serene. It was almost as if the swell moved in slow motion. And the northerly wind had, at last, brought reprieve from the tropical heat.

Jennifer took me aside to say she was worried about 'our' Benji, who was on a collision course with several people. Captain announced he'd brief us at 16:00 on the 'heavy weather conditions' we might expect in the Atlantic Ocean, hinting this was 'packaging,' a Trojan horse, to discuss social relations aboard. Despite his best intentions, Captain's talk missed the mark. As often happened, he waffled on when speaking in front of the group and failed to convey the message he so clearly managed to share with me earlier: we all needed our own time and space and pace; we had to respect this in both ourselves and others. Despite the growing tensions, we also started taking more time to speak with and listen to each other. Some conversations started to deepen, while others flattened by the day. One person I spoke to more often than before was the Bo'sun, who'd had a sheltered and devout childhood on an isolated family farm in West Virginia. Farm life made him a great Bo'sun, as he was clearly well versed in the art of having to do everything with nothing. Because he had joined Cornelius' project while the *Avontuur* was undergoing a major refit so she could carry cargo again, he knew the vessel inside out.

Most of all, I had a clear routine by this point. I got up around 03:30 in the morning, in time for watch handover at 04:00. I helmed the ship, trimmed sails, cleaned a bit, and looked at the rising sun, until we gave the deck to the Red Watch at 08:00. After handover, I had breakfast rather quickly and went to bed to watch a movie. After that, I napped until lunchtime at noon. I tried to stay awake after lunch and speak to people. But for lack of shade, or something new to say, I'd often retire to my bunk and sleep some more. At 16:00, I was on watch again. I had become increasingly solitary. I did not mind.

I found Geraldine Brooks' *Year of Wonders* aboard, which recounts the struggles of the villagers of Eyam in 1666, when the black death ravaged England. As I read it, the deeply religious calls for repentance in the face of the invisible force of the black death made me wonder: what did this say about the pandemic we were facing? Captain, like the rector in Brook's fictionalised history, implored us to welcome the challenge we faced for personal growth. Perhaps ironically, as everyone ashore was masked, Captain argued that we on the *Avontuur* no longer had to live with masks. We no longer lived out the masquerade that is social life, for we could no longer keep our masks on. We had nowhere to hide. That is the mantra we all kept repeating throughout the voyage. There is no hiding. Most of all, there is no point hiding, because everything is out in the open.

I'm not sure this was true. I, for one, remained far more polite and civil than I wanted to be at times. I kept up a façade to both hide myself and remain respectful to others. I can scarcely believe that others did not do the same. We were not able to hide our actions, though. Others did not perceive us through our ideals and words, but through our everyday behaviour.

Despite a lifetime of sailing, Captain had never experienced anything like the voyage we were on. We may not have agreed to be there for the 'emotional intelligence field experiment' as he called it, but we were there nonetheless. He considered our adventure a unique opportunity to get to know ourselves better. We all had our reasons to be here, he thought. However, he never volunteered to tell me what *he* learned from the experience, or how the voyage changed *him*.

Allan Sekula argues that large-scale shipping has transformed economies. As factories ply the oceans searching for cheaper labour, the ocean has become predictable and monotonous. Thanks to the uniform logic of the shipping container, the sea of 'exploit and adventure' turned into a 'lake of invisible drudgery.'[14] Is it mere drudgery? Or is the ocean, so carefully constructed as being somehow separate and distinct from land, no longer a mere conduit of trade?

'For a long period,' the anthropologist Ghassan Hage reflects, 'we humans have felt reasonably secure in our capacity to utilise nature and its resources. We felt that resources were either endless or able to regenerate themselves. And while we knew that there were all kind of toxic and polluting materials that resulted from both the extraction of resources and their consumption, we felt able to manage, recycle, or at least live with this waste.'[15]

Ocean water is absorbing CO_2, leading to acidification, which is driving the 'most rapid and greatest changes in ocean carbonate chemistry' in 'tens of millions of years.'[16] The chemical composition of seawater is changing so much that it is becoming inhospitable to many living creatures. Arctic sea ice is melting. Sea levels are rising. More than 90 percent of global warming has been absorbed by oceans.[17] The immediate effects are felt on land, and they increase the risk of 'positive feedback loops' that could propel climate change beyond dangerously critical tipping points.

Before setting sail, Cornelius told me that he found ocean currents increasingly unreliable. The trade winds on which sailing ships have relied since they ventured out to sea may be changing because the oceans are changing. Some fear that age-old water currents in the Atlantic Ocean have already undergone shifts that may indicate they're reaching or exceeding climactic tipping points.[18] The implications of changing wind and water patterns for wind propulsion remain unclear.

'The amount of global warming,' James Lovelock argues, 'depends hugely on the properties of water. When cold ice forms, much of it is white snow. This reflects the sunlight back to space and is cooling. But when it is warm, the water vapour in the air is a powerful greenhouse gas that makes it warmer still.'[19]

This is not merely an observation of natural phenomena. Searching for causes, Liam Campling and Alejandro Colás find them in the wilful exclusion of 'negative externalities' from corporate balance sheets.[20] 'The twin signatures of this era,' Naomi Klein argues, 'have been the mass export of products across vast distances (relentlessly burning carbon all the way), and the import of a uniquely wasteful model of production, consumption, and agriculture to every corner of the world (also based on the profligate burning of fossil fuels). Put differently, the liberation of world markets, a process powered by the liberation of unprecedented amounts of fossil fuels from the earth has dramatically sped up the same process that is liberating Arctic ice from existence.'[21]

None of this is particularly new. Grotius wrote his pamphlet in 1604, sixteen years before the unholy Dutch conglomerate of traders and mercenaries of the *Vereenigde Oostindische Compagnie* attempted to eradicate the entire human population of Lonthor, a Maluku island of the Banda group in present-day Indonesia. The company's murderous plunder served no other purpose than to establish a global monopoly in the nutmeg trade; to make money.[22]

'The present phase of the planetary crisis is not new at all,' argues Amitav Ghosh in *The Nutmeg's Curse*, 'rather, it represents the Earth's response to the globalisation of the ecological transformations that were set in motion by the European colonisation of much of the world.'[23]

If this sounds far-fetched, it's worth bearing in mind that even IPCC reports now acknowledge the connection between colonialism and climate change.[24] For example, colonial trade, as we've seen between Europe, Africa, and the Americas, relied on the triangular trade in manufactured goods (towards the colonies), slaves (towards plantations in the 'New World'), and produce (towards Europe). What happened was not solely the inhumane treatment of both enslaved people and the extermination of indigenous peoples. This trade also allowed Europe to expand its population while increasing its consumption per capita, by extracting food, resources, and wealth from colonies of exploitation.

'If one country exchanges its labour for the raw materials of another country,' claims Aleksandr Ėtkind, in summarising the eighteenth-century *Essay on Economic Theory* by Richard Cantillion, 'then the first country will have the advantage in this trade, since it maintains its people at the expense of that other country.'[25]

How does that work?

Today, these exchanges are complex, because they involve long and intricate webs of ownership, production, processing, and transport. Looking at early histories of regional trade gives an insight into how differences in wage costs, value added, and crop yields created significant imbalances of trade between regions. In the 'trade between Paris and Brussels,' Ėtkind continues by way of example, 'the harvest of a single acre of Flemish flax with value added by Brussels lace-makers, is equal in value to the wine from sixteen thousand acres of French vines.' This has important implications for the use of space, as 'many people could live on every acre of Flemish land, doing other things … and buying food from Burgundy or elsewhere.'[26] It was, simply put, the import of food from far-flung countries that allowed Europe to prosper.

The globalised economy is a whole lot more complex than Ėtkind's example of the historical trade between my native Flanders and Burgundy suggests, in large part because transport has become so cheap. And that's mostly thanks to Grotius' fiction that maritime transport would not harm anyone. The mere existence of ships did not kill the population of Lonthor or other colonies. But without ships, the Dutch would not have been able to trade over such distances. Hence, without the free-for-all Grotius imagined and encouraged, world trade would have been very different.

The use of fossil fuels for propulsion of cargo ships now causes significant harm to the planet, by adding huge amounts of greenhouse gas emissions to the atmosphere. Shipping executives keep on peddling the fiction that it is a mere conduit of trade. But it is shipping that makes trade possible. By excluding questions about the shape and size of shipping from ethical considerations about the planet, we're continuing to justify the belief that the world is ours to take.

Governing the ocean

The International Maritime Organization (IMO) looks like any other UN agency to most people. It has a hundred and seventy-five member states, which technically each have one vote.[27] In principle, the Marshall Islands with its 58,791 inhabitants has a vote equal to that of far more populous countries like India or China, with 1.366 billion and 1.398 billion people respectively. This aligns with how the United Nations work in theory. Though few UN agencies vote on issues; they rather reach consensus, meaning *unanimous agreement* on a final text. The IMO has made a similar shift, though it aims for *broad agreement*, not unanimity.

At the IMO, there is also an imbalance of power, but it manifests rather differently, with consensus-driven decision-making favouring countries with significant interests in maritime transport. This is because of the composition of its executive body, the Council consisting of forty member states. States are elected to the council. But for the two-year terms of council, members do not result from open elections in which each country has the same opportunity to present its candidacy. Membership is based on three categories. Category A is reserved for '10 States with the largest interest in providing international shipping services,' Category B for '10 States with the largest interest in international seaborne trade,' and Category C for '20 States … which have special interests in maritime transport or navigation and whose election to the Council will ensure the representation of all major geographic areas of the world.'[28]

To make matters worse, countries often 'grant shipping registries a formal position as delegates to represent them at IMO' or 'bring corporate officials from shipping registries as advisors.'[29] Such shipping registries are the bodies that register ships under a national flag. In many cases, they are government bodies, beholden to their responsible minister, but in the case of flags of convenience, there is often a disconnect between registry and government, as the former is often a privately owned company that operates on a profit-sharing basis with the state it represents, without political oversight or accountability. The registry of the Republic of the Marshall Islands, for example, is run by International Registries, Inc., which is based in Reston, Virginia USA.[30]

The Kyoto Protocol tasked the IMO with setting up a framework to regulate international shipping emissions in 1997. But rather than heeding this call, the IMO dragged its feet for twenty years. Not until 2018 did it come up with a clear industry-wide target to reduce greenhouse gas emissions. It will take until 2023 for the IMO to move beyond this 'initial strategy' and adopt a 'revised strategy' that should set out clear guidelines for member states to articulate domestic policies. Its decisions are slow to materialise, often lacking both urgency and ambition, but when they do the consensus-driven decision-making processes means that member states ordinarily implement IMO decisions into their domestic laws. While the power to set rules for ships resides with the countries they operate between ('port states') and the countries they are registered in ('flag states'), the IMO manages to set rules that have to be adopted into domestic laws once they have been agreed upon. In that sense, the IMO can be more effective than many other UN regulatory bodies. The weakest link is not the governance architecture of the UN agency, but the combined corporate influence and political obstruction that causes a 'race to the middle' that avoids the worst excesses, but fails to embrace the ambition needed to tackle climate change.

International shipping emissions occur between countries. So, they do not currently count towards national carbon budgets of either 'port' or 'flag' states, making it difficult to convince, let alone coerce, countries to take action. Furthermore, the deliberate opacity of the IMO makes it difficult to discern how decisions are made and whose interests they serve.[31]

This tension became all the more apparent when the IMO, at the 75th session of its Marine Environment Protection Committee in November 2020, translated its overall emissions reduction target set in 2018 into guidelines that will see aggregate emissions increase during the 2020s.[32] Simon Bullock, a researcher of shipping and climate change, and his colleagues argue that despite significant levels of 'committed' emissions (locked into the existing fleet), 'a combination of policies on low-carbon ships from 2030, combined with speed and operational measures from the early 2020s, could keep shipping within a Paris-compatible carbon budget.' This would, however, require more ambition than current IMO targets display.[33] Indeed, more stringent targets are necessary, and the longer it takes to set them, the bigger the challenge will be.[34]

The precise targets are not the only challenge. The IMO lacks a clear implementation pathway. The way in which it has translated its 2018 targets into operational measures will not see emissions drop. With such excessive delays and weak targets, it is hardly surprising that environmental activists have voiced concern about the IMO's ability to act swiftly and decisively.

On top of this, the IMO lacks accountability rules for its delegates in two key ways. On the one hand, journalists are forbidden from reporting the positions of individual representatives without their direct consent. This hampers the possibility of holding member states accountable, because meeting reports do not clarify the positions taken by representatives. This means that the public has to believe that member states vote according to their publicly proclaimed positions. It is impossible to know the positions taken and votes cast by member states. On the other hand, as the IMO has no rules on secondary employment or conflict of interest for delegates, a national delegation can include a representative from that country's shipping industry, who can speak on behalf of the country. Given the secrecy allowed to representatives, this means that industry representatives can weigh in on consensus-driven debates without any public accountability. When it comes to the environmental regulation of the shipping industry, this means that the public interest has little chance of gaining the upper hand against the short-term commercial interests of the industry.[35]

Given the lack of transparency of IMO debates, it is difficult to point the blame at any member state. But one recent example is telling. At the first One Planet Summit, an initiative of French President Emmanuel Macron, held in December 2017 at La Seine Musicale, a performing arts centre in the Parisian suburb of Boulogne-Billancourt, thirty-five heads of state signed the Tony de Brum Declaration. This signalled their commitment to ensure 'a peak on [shipping] emissions in the short-term and then reducing them to neutrality towards the second half of this century.'[36] The declaration was named after the late Marshallese politician who spoke out strongly for ambitious climate targets, as failure to do so would see his native Marshall Islands increasingly submerged under rising seas.

The first blow to this declaration came a year later, in 2018, when the IMO agreed on its 'initial strategy' in which the mid-century target shifted from 'neutrality,' a 'net-zero' ambition, to a mere minimum 50 percent reduction of greenhouse gas emissions. While it remains open to doing better, the IMO's target now lacks the ambition needed to meet Paris Agreement commitments.

The second blow came in 2020, when the IMO approved weak changes to the regulation of fuel efficiency in individual ships, which will likely see emissions grow by 14 percent before the end of the decade.[37] Despite commitments reflected in the Tony de Brum Declaration and the Paris Agreements, IMO member states did not heed Marshallese demands for a carbon pricing mechanism.[38]

Considering the plans and strategies that are on the table now, the IMO is unlikely to deliver the leadership that will bring down carbon emissions from shipping quickly and rapidly enough. This has long been known by politicians, campaigners, and industry leaders who pursue ambitious action. While the IMO has failed to take meaningful action, the European Union set stricter regulations through its Fit-for-55 package in 2021.

As the IMO favours technical debates over ambitious structural change, the organisation's legitimacy is eroded. Thomas Weiss argues that the United Nations as a whole is in crisis, because the predominance of state sovereignty hampers concerted action that takes agreements beyond narrowly defined national interests. Combined with the atomisation of mandates within the United Nations, this has dispersed and diluted responsibility, and the tensions between North and South hamper meaningful action on global issues, which results in meek and lacklustre leadership, leaving insufficient space for the leadership of agencies to drive bold action.[39]

In August 2021, the IPCC published the main findings of its sixth assessment report. The two main takeaways were that we need to attain net-zero by 2050, and probably sooner than that if we want to stick to the 1.5 degree target[40] – though all that depends on just how quickly we take action now.

'Climate change represents,' Amitav Ghosh reminds us, 'in its very nature, an unresolvable problem for modern nations.'[41] Nathaniel Rich explains why in his *Losing Earth: The Decade We Could Have Stopped Climate Change*, his depressing history of why climate action did not materialise in the 1980s, the decade in which I was born.[42] He argues it is a problem that cannot really be solved, at least not in a way that allows for a nice photo op, where politicians can claim victory.[43]

That's why we ended up with 'net-zero.'

Current commitments rely heavily on drawdown and capture. Its unproven promise risks slowing down emissions reduction efforts. The problem, moreover, is that it allows us to discount future risk even more than we have so far. Why suffer by taking action now, if the future brings zero-emission technologies – and technologies that promise to undo past emissions?

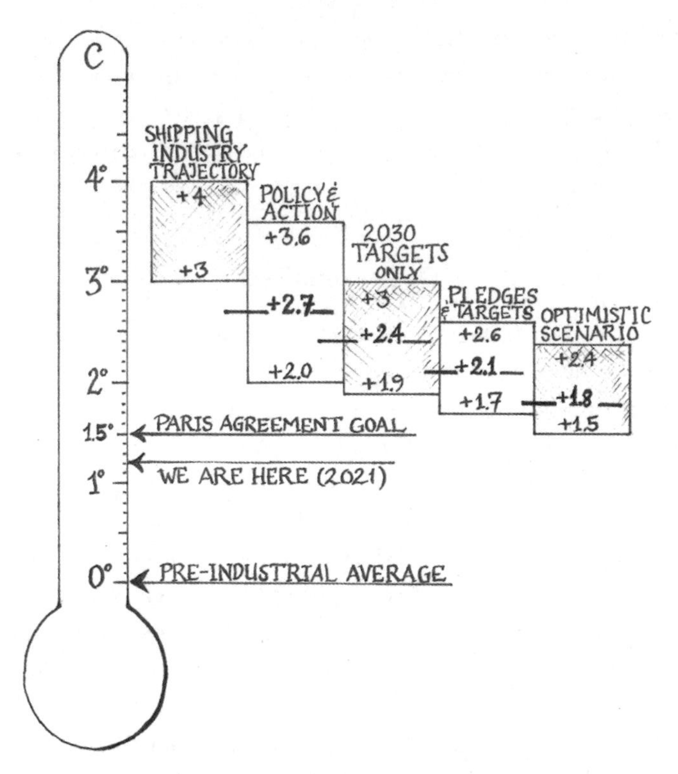

The *Climate Action Tracker* thermometer indicates expected temperature increases by the year 2100

This leaves us with the reality. Current levels of commitment in the shipping industry are well below 'net-zero.' Meanwhile, rich countries need to decarbonise faster than the rest of the world, as they can better carry the financial burden of the transition.[44] The combination of these two challenges makes net-zero a dangerous trap: on paper it looks like we're almost there, but will we be able to overcome all the hurdles that lie ahead?[45]

Even if demand for shipping diminishes significantly, the shipping industry requires rapid deployment of zero-emission technologies and fuels to meet the targets set out in the Paris Agreement. How will this happen if both the deployment of existing technologies and further innovation, as well as slow steaming, remain tenuous, given the lack of economic incentives (carbon price) and political pressure (regulation) to do so?[46]

In this difficult regulatory environment, can 'zero-emission' ships offer a way out?

Ocean rebellion

On November 16, 2020 a group of Ocean Rebellion protesters, the maritime spin-off of the environmental action group Extinction Rebellion, once again raised alarm in front of the IMO headquarters in London. Representatives of IMO member states were negotiating the scope of new environmental regulations for the shipping industry. That few IMO delegates were in London because the UN agency conducted its discussions entirely online, did not deter the protesters. Sophie Miller, a spokesperson for Ocean Rebellion, claimed that the proposed measures – which were adopted – 'will allow shipping emissions to increase rather than decrease.'[47]

Throughout the meeting, I followed the negotiations as best I could. Back home in Australia, I had started working on this book, living about as far away from London as I could possibly be. Even if I had been closer, it would not have helped much. The IMO is rather secretive and its decision-making processes deliberately opaque, as we saw in an earlier section. Journalists cannot freely report on negotiations, as they need to obtain express consent of speakers before naming them. And the precise positions of countries is not communicated. Only the proposals and decisions see the light of day. Between the ominous institutional silence and the bouts of protest covered by the news media, I dreaded the outcomes.

It is not easy for Ocean Rebellion to challenge opaque negotiations about invisible pollution that has long-term effects. Nathaniel Rich rightly asks how to take action in such a case: 'How do you stage a protest when the toxic waste dump is the entire planet, or worse, its invisible atmosphere?'[48] Even if protesters gather, their message is easily defeated through lack of attention, or through ridicule.

What was at stake at the IMO meeting in November 2020 was nothing less than the first real opportunity to turn the carbon emissions reduction target (halving emissions by 2050) into practice.

Did they succeed? Anything but.

'All we're saying,' Clive Russell of Ocean Rebellion told me in November 2021, 'is the UN needs to be more robust and that's not going to happen. That's really not going to happen. We know it's not. The UN is what the UN is, and it's not fit for purpose. I think we can't rely on the UN to save the ocean.'

When calling for a different kind of future, the challenge becomes far more complex. Building an appealing alternative is far more difficult than cutting carbon from our economies. Expecting the United Nations to lead in that regard would be tantamount to asking them to shift the ideological orientation of the global economy. No one, certainly not Ocean Rebellion, expects them to do that. Nor did anyone mandate the organisation to do any such thing.

'One of the reasons why non-violent direct action is important,' Russell continued, 'is because it brings people together and empowers them.' It allows people with diverging ideas and future plans to come together and explore overlaps and tensions between their ideas. It is precisely those collective actions that function as 'stepping stones to get people to realise that they can be more powerful than they think they can be.' Most importantly, in my view, it acknowledges that there is no simple solution to the intertwined set of problems we face.

'Lots of people think there's a silver bullet to this situation where there really isn't,' he went on. Big guiding ideas can help rally people together in a struggle, though ideas alone don't solve problems. The anthropologist Anna Tsing describes the tension that emerges between 'a global dream' and the place we dream from. She uses the metaphor of a bridge to describe the challenge of putting big ideas into practice. The bridge, we hope, will allow us to cross from the 'parochial island' that is our lived situation to the abstract ideals and futures we envisage. This tension does not quite manifest until we try to change our local reality to the abstract global ideal. In crossing the 'bridge of universal truths [that] promises to take us there,' we don't arrive at our imagined destination, because 'we find ourselves, not everywhere, but somewhere in particular.' This tension between an imagined future and where we actually end up is what she calls 'friction.' There is indeed friction between the bridge we stepped on and the bridge we stepped off. 'It is only,' Tsing argues, 'in maintaining the friction between the two subjectively experienced bridges, the friction between aspiration and practical achievement, that a critical analysis of global connection is possible.'[49]

The very idea of friction is precisely what makes bringing people together so vital to climate action. On our own, we can imagine having found a neat solution. To return to Clive Russell, 'if you think that solutions reside within yourself, that's problematic because I don't think they ever do.' What is more, even if you found the solution, it could never come to fruition as imagined, because one person alone cannot possibly accomplish the fundamental shifts the climate crisis needs. 'There's no Jesus Christ in this story, unfortunately.'

But bringing people together, 'collaborating, listening, working with lots of different people,' is what Clive Russell thinks 'really helps.' It does for Ocean Rebellion as much as it does when sailing together.

'Mission zero'

On May 22, two-and-a-half weeks after leaving Veracruz, Bo'sun built a cooking contraption in one of the galley benches. It was a combination of woven glass fibre fabric and expanding insulating foam. It fitted neatly around two cooking pots, allowing Cook to bring soup or stew to a boil and keep the food simmering for hours without requiring further energy. It was a practical and functional way of saving gas, which was running frighteningly low, as the bottle refill in Veracruz was more a half-refill. Bo'sun's nifty innovation only entered service long after we reached 'peak gas' on board. Much like the rest of the planet, we did not start implementing practical, feasible, and effective ways of reducing our fossil fuel use until it had almost run out.

Timbercoast's 'mission zero' involves a five-state approach that starts with raising 'awareness about the environmental destruction caused by the shipping industry.' They respond to this by modelling 'a clean shipping future' with the *Avontuur*. They do so by running a commercial enterprise, which sells 'premium *Avontuur* products to support the ongoing operation of the project.' This, they hope, will 'establish demand for products shipped by sail,' which would allow them to eventually 'build a modern sail cargo fleet.'[50]

The aim was never to just sail a tiny schooner to compete with container ships. This echoes the ideas of Fairtransport since they started sailing the *Tres Hombres* in 2007:

> Fairtransport's 'Mission Under Sail' is to minimize the carbon footprint worldwide and to raise awareness for climate-friendly transportation. We strive for a sustainable product lifeline and aspire to constantly improve our own environmental impact. Since 2007, our positive community is trading with local entities and moving organically grown goods from fair producers to conscientious consumers. The long-term mission of Fairtransport is to revive cargo shipping under sail, to build an impactful and strong worldwide movement, and to inspire others with alternative shipping methods.[51]

For Sail Cargo Inc., the company building *Ceiba* in Costa Rica, the 'mission is to prove the value of clean shipping,' as 'developing a resilient, decarbonized transportation sector is one of the most critical tasks of the Anthropocene.' Like Timbercoast, the way in which they plan on doing this favours a combination of old and new technology: 'Eliminating fossil fuels from the maritime sector is achievable by using advanced technology and simple techniques. We combine innovative, clean technologies with readily available, low-cost, natural systems to create solutions for a range of needs: from supporting vulnerable coastal communities to moving commodities at global scale.'[52]

But what makes a vessel zero-emissions?

While we sailed the *Avontuur*, we lived in constant friction between mission and reality, as Anna Tsing calls it. The ship combined the ancient zero-emission technology of sail with high-tech navigation. At the same time, we relied on the electricity-driven water maker to keep us alive. On average, we used some fifteen litres of water per person per day. The water maker could run non-stop and generate far more water – up to around one hundred litres per person a day. At sixty-five litres of water per hour, Captain told us the water maker 'costs' about one litre of diesel per fifteen litres of water. When I asked him about this later, Cornelius disagreed, by pointing out that while the generator does use some three and a half litres of diesel per hour, most of the electricity produced is needed to top up the batteries on board. So Cornelius thinks that only half of that fuel would be used by the water maker, meaning that one litre of diesel 'makes' about thirty-five litres of water. By design, the solar panels and wind turbines should generate enough electricity to power all equipment aboard. Sadly the wind turbines weren't working. And the planned generator that would harvest the kinetic energy of the propeller shaft was yet to be installed.

With insufficient solar and wind energy, we needed the generator to provide enough electricity for the water maker and navigation equipment. In port, we ran the diesel engine for manoeuvring. And on occasion, we motor-sailed out of high-pressure areas.

So, we were on a mission to zero emissions – though we were not quite there yet. But just how high were our emissions?

To answer this, I started by establishing the distance from the first port where we picked up our total of 65 tonnes of cargo to Hamburg: 8482 nautical miles. We used 2468 litres of fuel on the voyage. (I have included the fuel use of the entire voyage in my calculations, because the westbound voyage carried no cargo to speak of.)

Bearing in mind that every litre of diesel results in 2640 grams of carbon dioxide, I calculated that our total emissions were 6.52 tonnes of carbon dioxide. That was a tad more than my round-trip economy flight from Australia to Europe which added up to about five tonnes. Considering there were fifteen of us, carrying sixty-five tonnes of cargo, that's not all that bad.

But how much carbon did we emit per tonne-mile? This is the measure by which we're able to compare the carbon emissions between different types of transport. It gives, in grams, the amount of carbon we emit to transport one tonne of cargo over one nautical mile.

As we shipped 65 tonnes over 8482 nautical miles, the total tonne-miles were 551,330. Considering that the shipping industry, as a whole, tallies up to more than sixty trillion tonne-miles a year, one hundred and ten million times more, that's small fry. But to get to the crucial number, we divide the total carbon emissions by the total tonne-miles. The result?

11.82 grams of carbon dioxide per tonne-mile. That's hardly zero emissions.

How does that compare with the rest of the shipping industry?

In 2018, the average emissions were 11.67 grams of carbon dioxide for every tonne-mile, down from 17.1 grams in 2008.[53]

In our defence, much of our motor-sailing was in response to tight port schedules because of the pandemic – as well as motoring to meet the Spanish Coast Guard after taking sixteen refugees on board. More generally, I do not include this calculation to make the case that lowering emissions by shipping cargo under sail is impossible. I merely aim to show how difficult it is to beat the shipping industry on carbon intensity. Operating at such a small scale means that even propelling the ship by diesel engine for a measly 2 percent of the time undoes the obvious potential of wind. 2020 was, indeed, an exceptional year in many respects, including the amount of fuel on our Atlantic round trip. Needless to say, the fuel used by small ships like the *Avontuur* is far less polluting than Heavy Fuel Oil, which is toxic sludge.

In 2021, Cornelius told me he 'sailed right from alongside Tenerife wharf until dropping the anchor [in Marie Galante] without ever touching the engine! Just above 3000nm [nautical miles] with zero fuel for the main engine.' Even so, they had to charge their batteries and make water by running the generator, which led them to burn 'close to 100 litres of Diesel.'

With the tiny cargo loads the *Avontuur* can carry, even those small amounts add up. It is, after all, hugely challenging to compete with an industry that has attained such economies of scale. On top of burning diesel, our sails and ropes are made from plastics, which rely on the extraction, refining, use, and eventual discarding of fossil fuels. The ship is thus a contradictory hybrid that simultaneously embraces and rejects technology and craft. We could, after all, navigate by sextant and paper charts, while hoisting linen sails up the masts using manila ropes.

Even when very pointedly aiming for zero-emission sailing, challenges remain. In preparation for the build of the new *EcoClipper*, Jorne Langelaan commissioned a report on the lifecycle emissions of the vessel. The verdict? She'll emit some 2 grams of carbon dioxide for every tonne-kilometre.[54] Not zero, but a lot closer. Most of the *EcoClipper*'s expected emissions will relate to the build and maintenance of the vessel. I did not count these when calculating the emissions of our *Avontuur* voyage. Then again, no shipping company calculates these lifecycle costs. Their focus is on the operational emissions when burning fuel to move their vessels – which remains by far their greatest source of emissions.

In a commentary on the *EcoClipper*'s lifecycle assessment, Kris De Decker, the author of the deep-green *Low Tech Magazine* (he runs the server on power generated by a balcony-mounted solar panel from his Barcelona apartment, to give you an idea), argues 'we should not be fooled by abstract relative measurements, which only serve to keep the focus on growth and efficiency.'[55] In other words: since attaining absolute zero emissions is impossible, we should probably question the growth trajectory of the global economy too.

On Sunday May 24, the Captain's Reception had an exciting backdrop: the coast of Cuba. Seeing land was the most exciting thing that had happened in weeks. The gloriously green hills and mountains rising into the mist beckoned to us. I would have enjoyed a stroll ashore, but our only contact with 'civilisation' was an empty upturned soft drink can floating past our vessel.

Shortly after, we spotted the first yacht we'd seen since leaving Pointe-à-Pitre at the end of March. The *Ganny Mede* from Los Angeles, California, was on her way to Fort Lauderdale in Florida.

In the afternoon of May 25, Little Captain was in his bunk during his watch. I wondered what had happened.

Mia soon told me that there would be a watch shuffle again. She was returning to the Green Watch. Pinkie would join us in the White Watch. And Little Captain would go back to the Red Watch, where Captain would be able to keep an eye on him. The Red Watch was soon to become, in Jennifer's words, 'a herd of bullocks,' in which three men, Benji, Martin, and Little Captain tried to prove something.

Captain kept on saying how he thought everyone on board had a reason for being there. Each of us, he said, had something to learn. I was starting to understand what he meant. We were there to understand the complexity of the challenge we faced.

'It is more than climate change,' the interdisciplinary scholar Donna Haraway writes. 'It's also extraordinary burdens of toxic chemistry, mining, depletion of lakes and rivers under and above ground, ecosystem simplification, vast genocides of people and other critters … that threaten major system collapse after major system collapse after major system collapse.'[56]

We were not there to simply rid shipping of carbon emissions. As necessary as it is to fully decarbonise the shipping industry, we can't rally the world behind a goal as abstract as 'net-zero' or even 'mission zero.' We need a story that takes us there.

Groundhog Day

When passing Miami on the night of May 26–27, it finally felt like we had turned a corner and started our return passage in earnest. Even so, we had no idea what was going on in the USA, within view, but worlds away. We did not know that George Floyd had been killed the day prior, nor would I learn about the ensuing Black Lives Matter protests until much later.

Much of our progress was thanks to the Florida Strait surface current of some five knots, which led us directly into the Gulf Stream. The wind also picked up. So much so, that Captain had started worrying about a 'disturbance' in the mid-Atlantic that could develop into a hurricane. While the season was yet to start – historically, it would start on the first of June – he was keen to ensure we stayed well clear of any rough weather.

We were not alone. Since mid-May, hundreds of yachties had started crossing the Atlantic. The pandemic saw them stranded for months on end in the Caribbean. Staying there with the hurricane season starting was not an appealing option, even when the alternative was crossing the Atlantic single or short-handed.[57]

Staying ahead of rough weather was not only important for our safety but played into several strategic considerations. First, we needed to arrive in Hamburg as soon as possible so Timbercoast could get paid for transporting cargo. We also needed to arrive to replenish *Avontuur*-branded coffee, rum, and chocolate, as Cornelius had run out of stock. Second, seeking refuge before we would reach Horta, our next port of call in the Azores, would cost money and time, which Timbercoast could not spare. Given the tensions on board, it would probably also mean losing some crew, which would make sailing onwards more difficult again. Third, Cornelius and Captain were on ever tenser terms. Captain prioritised safety and life aboard while Cornelius was, by this point, running the company on his own. His family life was suffering from Timbercoast's difficulty in turning a profit.

The suggestion that we might have to call at Freeport in the Bahamas if the 'disturbance' developed into a tropical depression did not go down well. I was not privy to the exact words Captain and Cornelius exchanged during a tense phone call, but they clearly did not see eye to eye. Captain insisted there was rough weather to watch out for – and possibly shelter from. But Cornelius saw no evidence of the peril, Captain told us.

Captain preferred playing it safe. With good reason. In 2006, while he was serving as master of the *Picton Castle*, a crew member was washed overboard and drowned during a fierce storm after sailing from Luneburg, Nova Scotia.[58] I too would prioritise safety over speed to avoid any such accident from happening again. It wasn't until Cornelius found out about this that he understood why Captain was so cautious.

On May 29, Captain told us that Cornelius had just 'admitted' that there was indeed bad weather in the mid-Atlantic. It was, by that point, classed as a 'tropical disturbance,' rather than the mere 'tropical low,' of two days prior.

We continued on a northerly course along the US coast. At some point, we would have to turn due east. Beyond concern for our safety, Captain told us that once we did so, we would not be able to turn back. Wind and current would be against us, making it hard to change course, apart from heading further and further north. And that would push us towards the Labrador Sea, where another challenge could await us: sea ice.

Meanwhile, the fight against entropy and hunger was endless. Especially surrounded by salt water, both rust and appetite seem to come back more quickly than ashore. Twice a day, just before lunch and dinner, someone would blow the conch to call hungry souls to the galley.

For breakfast, Cook served us porridge and bread. With fruit, if we had any. But by this point in the voyage, fresh food had all but run out. On occasion, we were treated to crêpes. As rationing tightened, we had less variety and fewer options. Bread-baking now happened only every other day. Condiments were running low. Unlike so many people ashore during the first lockdown, we gradually baked less sourdough bread as the voyage progressed.

Benji calling all hands for dinner by sounding the conch, April 3, 2020

The simple pleasures on board at this stage were simple indeed. The occasional bucket shower on deck brought some relief from heat, humidity, and sweat, though it did little to change the microclimate in my bunk, which felt like a pile of rags drenched in brine rather than anything remotely resembling a bed. Sadly, I could only blame myself for this. All I had to do was wash my sheets.

What I could not change was the gradual shift from Cook's sumptuous meals to a strict diet of beans with a side of beans. At first, she treated us to anything from *empañadas* to pasta to risotto to pizza. My favourite dish was undoubtedly her signature blend of lasagne and parmigiana. It was, effectively, a lasagne with rich layers of roasted aubergine. It manifestly pained her to work with ever-shrinking options. But it was the lack of fresh vegetables that made the tightening rations more difficult than the lack of variety.

Meanwhile, more and more things emerged about Little Captain's insubordination a few days prior. He had threatened to walk away from the watch as he disagreed with a manoeuvre.

To 'take back control,' he had secretly changed the course while at the helm. He thought he knew better than Captain or the Second Officer, his watch leader. He did so to 'start a revolution,' which on a ship sounds an awful lot like mutiny. On any other voyage, the master of the ship would have left Little Captain in the next port of call. Absent that option, we stuck together.

'I have no fancies about equality on board,' said Richard Henry Dana Jr. 'It is absolutely necessary that there should be one head and one voice, to control everything, and be responsible for everything. There are emergencies which require the instant exercise of extreme power.'[59]

As watch leader, the Second Officer could hardly do his job with a recalcitrant element on his watch. Beyond sneakily refusing to follow orders, Little Captain also switched to speaking German to fellow crew. This was both rather rude and against ship rules, as neither the Second Officer nor Athena spoke German. The working language on board had always been English, a practice everyone needed to uphold while non-German speakers were around. Cook and Jennifer would often speak in Italian. I would speak French with either of them, as would Athena. Merely speaking another language was never an issue, but doing so in the presence of someone who didn't understand always was.

Every last one of us on board made mistakes. Whether technical or social, we slipped up. But like Phil Connors in *Groundhog Day*, it was difficult to muster the energy to keep on tweaking life on board. We could not escape the unrelenting routine of the ship.

While travel is slow, social relations are fast-paced and highly volatile, even incendiary. Much like the seemingly endless ocean that consists of innumerable tiny constellations of water and air at different temperatures, the weather can shift quickly. The echo chamber of tiny incidents that get amplified and take on a mythical status in the collective awareness is central to shipboard life. Problems easily culminate into tipping points, which lead to conflicts that resolve as quickly as they appear; as it is impossible to leave, conflicts need to be quickly resolved or forgotten.

Some things, like the mould creeping up the hull of the ship, latching onto our every last possession, were difficult to keep at bay. Without a good dose of vinegar, we'd never get rid of it. And even with vinegar, it kept coming back.

Every day was, like Phil's, essentially the same. We were set in our ways and habits. But despite that recurrent sameness, our own minute choices made all the difference; perhaps the only difference.

7
Ship Earth

All is finite

Ships are self-contained vessels that float on water. So much for stating the obvious. I'm not the first to realise this. Every sailor setting out to sea knows they need to stow all the food and water they'll need to survive. Ellen MacArthur, a British sailor who broke the world record for the fastest solo circumnavigation of the world under sail in 2005, reflects on this challenging task:

> Your boat is your entire world, and what you take with you when you leave is all you have. If I said to you all now, 'Go off into Vancouver and find everything you will need for your survival for the next three months,' that's quite a task. That's food, fuel, clothes, even toilet roll and toothpaste. That's what we do, and when we leave, we manage it down to the last drop of diesel and the last packet of food. No experience in my life could have given me a

better understanding of the definition of the word 'finite.' What we have out there is all we have. There is no more.[1]

By early June, we on the *Avontuur* had finally started covering some distance northwards along the US east coast. The frustrating voyage out of the Gulf of Mexico had taken us far longer than we'd hoped. It had depleted our energy and our supplies. We were about to start our eastbound Atlantic Ocean crossing, but we had very little fresh food left. We had enough dry and canned food aboard to reach Horta, if we managed to sail on without further delay. Our remaining stores would not last long enough if we encountered adversity. Captain optimistically told us the 'fast train' – a fresh breeze that would whisk us across the Atlantic – was imminent. He sounded as if the Azores were nearly in sight.

When, a few days later, on June 5, my Spotify app stopped working because I had been offline for thirty days, I realised just how long we'd been at sea already. No more music to listen to in my bunk, and Horta was still a long way off. I was tired, irritated, and lacking motivation. I slept a lot, unable to focus on anything or open up very much. I was, in short, bad company.

The 'fast train' however, never really materialised. While we had some good days, most of the time was very slow going. The appeal of continuous, motorised propulsion started to dawn on me. Steam meant liberation from fickle winds; a prospect many a sailor must have appreciated. I was ready to chip in for fuel.

The glorification of slowness as wilful defiance of capitalist productivity never sounded so empty. Most of all because I was longing for fresh food. Rather than heading to the shops to replenish our dwindling supplies, we started rationing. Degrowth sounds nice in principle, doing more with less, but just how much would I want my own consumption to shrink? Admittedly, by joining a sailing ship, one's consumption becomes limited anyway. No more next day delivery. No more quick runs to the supermarket. No more options to buy new shorts (which I sorely needed by that point). And no more visits to book shops – which is arguably the only thing I missed. My consumption had indeed shrunk to a bare minimum, with even my basic needs controlled by a Cook on behalf of the shipboard community. This level of reduced consumption I could deal with, but the prospect of food rationing wasn't great.

When sailing, the wind always tells you what to do. And for the next forty-eight hours, we did nothing but bob up and down in the swell. Our calculations of just how much longer we could hold out without re-provisioning depended on how much we ate, as well as the one part of the voyage we could not control: the wind.

The next day, June 11, Athena woke me, saying we were doing six knots. By the time I was on deck, that had dropped to five. And during our watch, the wind abated, our speed diminished to less than three knots.

That day, Captain confirmed the whispers that we might head to Halifax for re-provisioning. The high-pressure area around the Azores would force us further north to avoid a mid-Atlantic calm. At that point, Halifax was about 365 nautical miles north-northwest of us. It would mean backtracking a little, spending more time at sea, postponing our arrival at Horta. All bets on when we'd reach Hamburg were off.

Worryingly, toilet paper was running low, too. Captain's suggestion that we could simply use water and soap – as he would – did not go down well with everyone. It reminded me of the time I tried to fight off Giardia at a bus station in Damascus, where no such luxuries were ever at hand. While at sea, I had no idea how pandemic-induced hoarding left supermarkets without toilet paper.

Cook did not respond very well to the news of a possible extra stop in Halifax. If we called there, she said she'd jump ship to be with her partner. This would cause Cornelius a massive headache, as finding a replacement cook would be difficult at such short notice.

Little Captain disliked the prospect even more. He was, after all, one of the few people on board for whom the extended voyage was causing real material harm. His company back home was struggling because of the pandemic. Being away on the *Avontuur* was costing him money. But as he pondered the threats to his business, an existential danger hung over us: without restocking, we risked getting stranded in the middle of the ocean without food.

At lunch, Captain outlined the severity of our situation. We had about twenty-one days of food left, and some twenty-five days of diesel to run the water maker for several hours a day. We probably needed much, if not all, of the diesel as venturing further north

The North Atlantic Ocean, June 19, 2020

would leave us with less intense sunlight to generate electricity, despite the longer days of summer.

We did not have to make a decision just yet. We would wait until we were directly south of Newfoundland. If we did call at Halifax, we would re-provision and leave right away, as Captain stressed that the hurricane threat was real, even this far north.

We kept sailing north. At 04:00 the next day, June 12, the sea temperature was still a balmy 23°C, but by 06:00, it had dropped to 14°C. It looked like we'd crossed the northeast-bound warm Gulf Stream into the Labrador current. But we were not quite there. We kept heading north and at roughly the latitude of New York City the sea temperature was back around 21°C. We still needed to sail a good bit further north before turning eastward.

On June 14, I spoke to Captain after lunch about the course and our collective plans. At the end I said, 'Let's hope the weather will allow us to make some distance soon.'

'I do not hope,' he replied, falling back into his tired old explanation. 'Hope is for beginners.' Captain either 'does' or 'doesn't' do things.

I must have said that I'd convince him that I'm right about saying 'trying' and 'hoping.' Turns out he kept that in mind throughout the voyage. After much to-ing and fro-ing, he sternly told me that he did not like me 'imposing' a certain way of speaking – or rather, thinking – about things.

He seemed to forget that, as the master of the ship, he was the one who was imposing views. Not by explaining what he meant, but by insisting that we speak in a certain manner. As I explained earlier, he insisted we do not speak of 'problems' aboard.

I strongly disagreed. I believe we're always merely 'trying' to make things work. We solve 'problems.' And we 'hope' that things will be fine. Not for lack of motivation or effort, but because of the sheer uncertainty we face in life. Particularly when sailing.

Deep down, I think Captain and I would agree. I simply aimed to clarify what I meant. He'd stress the importance of intercultural sensitivity and mutual respect, though he did not seem to think that such difference was manifested through language.

Perhaps he had taken what I said at the start of the voyage as a real threat. He did not ask me to explain what I meant at the time, but drew his conclusions and stuck with them. Throughout the voyage, I playfully referred to our semantic differences – though I never meant to suggest I was *trying* to make him change his mind.

Perhaps I saw my own stubbornness, pedantry, and tightly guarded fortress of thoughts and feelings reflected in Captain. I also keep up a shield. Confronted with criticism, I also deflect and hide behind linguistic technicalities. I am a pedant, nit-picking on all aspects of language and meaning. So, who am I to blame him for being similar?

Later that day, June 14, we started a 'trial' period of food rationing. We wanted to see how well we, individually and collectively, would cope with smaller portions.

Once we turned east, we would not be able to simply turn west again; the currents and wind would be against us. Knowing whether or not we'd cope was crucial in our collective considerations of whether we'd decide to sail beyond the 'point of no return' towards Horta without additional provisions.

Captain asked Cook to prepare about a third less food. This meant no seconds and no nightly snacks. This may hardly sound like rationing. It wasn't. But it confronted us with the finitude of our tiny world. Meanwhile, temperatures dropped significantly and we – or I, at least – needed extra fuel to keep warm during my watch.

The next day, June 15, I spotted sperm whales, a moonfish, and a school of dolphins. But one night, not much later, I made a cardinal mistake by staying in bed. Whales were playing at the bow, lightened by bioluminescence. I thought they were exaggerating when they called me from my sleep. So often the whales we saw were gone as quickly as they appeared. And more often, we'd only glimpse a sliver in the distance. That night, it was different.

And I was wrong; I missed the most memorable wildlife encounter of the voyage because I was cold and tired.

It remains hard to imagine just how much life there is beneath the surface, beyond the cetaceans that come up for air. While Robert MacFarlane's *Underland* reveals life underneath land surfaces, I had no way to discover the underwater world that is part of the complex web of life that is our planet.[2] Seeing the ocean surface mirror the light above it, I started understanding why we still know precious little about life beneath the surface.

With some 1200 nautical miles to go, Cook told us we had about two weeks of eating a 'varied diet' and about another week of more 'monotonous' food. Beans with beans, mostly.

Our world, the *Avontuur*, was a self-contained island as long as we were at sea. It was home, but not permanently so.

Common but undifferentiated?

'At exactly the time,' the novelist Amitav Ghosh points out, 'when it has become clear that global warming is in every sense a collective predicament, humanity finds itself in the thrall of a dominant culture in which the idea of the collective has been exiled from politics, economics, and literature alike.'[3] Naomi Oreskes and Erik Conway argue that the reason for this is the dominance of market fundamentalism in political and economic thought. This belief assumes that society's needs are 'served most efficiently in a free market economic system' and free markets are 'the *only* manner of doing so' in a way that does not 'threaten personal freedom.'[4] At the heart of this fundamentalism lies the belief that growth is both necessary and inherently good. Indeed, Tim Jackson argues it's the defining myth of our time: 'Every culture, every society, clings to a myth by which it lives. Ours is the myth of growth. For as long as the economy continues to expand, we feel assured that life is getting better.'[5] As a result of globalisation, 'our' cultural belief has reached virtually every society throughout the planet.

The result is that some organisations and industries, shipping and aviation more than most, have long focused on 'actively doing nothing' about climate change.

The pretence of being 'international' exempts these industries from fuel taxes and carbon targets. By tinkering with fuel efficiency, they hoped the problem would go away. And to some extent it has; the carbon intensity of shipping has reduced significantly from its frenzied high in 2008, just before the global financial crisis, when global trade demanded ships go at high speeds that were burning far more fuel per tonne-mile than they do now. The growth of shipping demand is, however, so high that total shipping emissions have continued to grow, albeit at a slower pace. This is why the IMO revised its projected carbon emissions downwards in its fourth edition of the Greenhouse Gas Study in 2020, compared to the third edition in 2015.[6]

In 2015, Koji Sekimizu, then Secretary-General of the IMO, opposed any change to the growth-oriented political economy of the shipping industry.[7] While stressing the potential to lower its carbon intensity by 30 percent, he doubled down on the need for economic growth and shipping industry growth – even if the IMO's own Greenhouse Gas Report at the time suggested that the total emissions of the industry would likely increase by between 50 percent and 250 percent by the year 2050.[8]

What changed? Ambition. High ambition.

Corporations and politicians long feared it would be (economically) impossible to decarbonise shipping. Pushing the industry to cut its emissions would threaten global supply chains, consumer choice, and ultimately corporate profits. While environmentally necessary, this was not a politically expedient or commercially attractive option, which is why they pulled every trick in the book to slow down action. Like true 'merchants of doubt,' industry executives called for more time to get to the bottom of the issue.[9]

This changed around 2018, when the Shipping High Ambition Coalition managed to get the IMO to commit to a decarbonisation strategy. The shipping industry would finally start taking action on climate change.

One afternoon, while he was at the helm, Bo'sun told me he thought that Cornelius' otherwise laudable insistence on playing by the rules was stifling the company. The cost of compliance was excessive compared to the potential. The main aim of ships like the *Avontuur*, as Bo'sun sees it, is to show that an alternative to fossil-fuel driven shipping exists. Raising awareness is crucial. Trying to bend German merchant marine regulations to accommodate a tiny century-old sailing vessel like the *Avontuur*, however, is unrealistic.

Before sailing aboard the *Tres Hombres*, Bo'sun thought of her owners and crews as 'crazy hippies who broke all the rules.' He has since realised they only break certain rules, and only after careful consideration. They may seem crazy, but the owners of the *Tres Hombres* are still in business after more than a decade, and they have been turning a profit for most of that time. The more he's involved in sail cargo, the more Bo'sun thinks the radical projects make sense. There's only so much you can accomplish as a tiny company. But the strength of the *Avontuur* or the *Tres Hombres* lies in their visibility, not their power as industry lobbyists.

In 2020, the *Avontuur* sailed under a German flag but is now registered in the Seychelles. As a ship is not required to be flagged in the country of its owner, many ships fly the flags of countries that have less stringent regulations. This is so common that the three largest shipping registries – Panama, Liberia, and the Marshall Islands – account for 42 percent of all registered tonnage.[10]

'International conventions that can actually be enforced and flag states that can be compelled into good behaviour' are crucial to undoing the harm of decades of legal loopholes created by flags of convenience, Rose George argues. 'Oversight, governance, fair play. All the things expected on land that dissolve twelve miles out, in open seas.'[11]

Flying a 'flag of convenience' allows shipowners to pay lower wages. Flagging ships in places with more 'convenient' regulations extends the logic of footloose capitalism, where factories now move to the cheapest and least regulated locale. This is only possible thanks to cheap and abundant maritime transport.[12] Shipping is not simply the handmaiden of trade, but the engine that drives capitalist expansion which is wrecking the planet. Shipping is, indeed, not a mere service; it is a vital node in the infrastructure without which global consumer capitalism could not exist.

The rules about flagging ships are so lax and easy that a ship can change flags overnight. This challenges the principle of 'common but differentiated responsibilities and respective capabilities,' which dominated thinking in global climate regulation. Poor countries, the reasoning goes, are less responsible for historical and contemporary emissions, so they should not have to cut their emissions as much as rich ones. This puts the IMO in a tricky position.

'The Earth is our shared roof and our shared shelter,' Achille Mbembe reminds us, just in case anyone might have forgotten. 'Sharing this roof and shelter is the great condition for the sustainability of all life on Earth. We have to share it as equitably as possible.'[13]

But what does that mean when the richest 10 percent of people, including me, are responsible for fully 52 percent of cumulative carbon emissions?[14] It is indeed the world's richest, with their private jets, superyachts, and mansions – but also with their polluting investments – who carry a disproportionally large share of responsibility for the climate crisis we're in.[15]

It means that we are indeed facing a common responsibility, but we don't share the same responsibilities. When on land, this kind of makes sense. But at sea, following this principle gets more difficult.

The IMO does not follow the 'common but differentiated' principle. This is not because the IMO does not recognise the unequal historical contributions of developed and developing countries, but because ships can change ownership or registry overnight. As a result, the IMO does not set different rules for 'developing' countries. There are, however, proposals, notably from the Marshall Islands (which is a country that issues flags of convenience through its open shipping registry, but also a country with high climate ambitions, and a country faced with an existential risk due to rising seas) and the Solomon Islands, to ensure that a possible carbon levy for the shipping industry will serve to compensate climate-vulnerable developing countries for the challenges they face.[16]

High Ambition Coalition

On April 6, 2018, just before IMO member states agreed on an 'initial strategy' to curb greenhouse gas emissions from shipping, Hilda Heine and Christiana Figueres wrote an op-ed for the *New York Times*:

> The Marshall Islands may be a tiny, climate-vulnerable nation of low-lying islands and atolls threatened by rising seas, but it also hosts the world's second-largest shipping registry and is almost entirely reliant on sea transportation for food and other crucial supplies. Given all this, perhaps no country is better placed to highlight the need to act, and to do so in a way that is economically sustainable.[17]

Heine was, at that time, the President of the Marshall Islands, while Figueres had finished her role as the Executive Secretary of the United Nations Framework Convention on Climate Change (UNFCCC, whose signatory states meet every year – hence Conference of the Parties), which she held between July 2010 and July 2016.

In their joint statement, they stressed that 'countries truly committed to climate action will not accept anything less than a strong deal.'[18] This was not the first case of climate leadership coming from the tiny Pacific state.

The diplomatic deadlock at the Copenhagen COP in 2009 left many, me included, wholly disillusioned that any real agreement to follow the Kyoto Protocol could be reached. The world needed both greater ambition in terms of targets and a geopolitical shift that would fairly share the burden of climate action, beyond the stifling opposition between 'developed' and 'developing' countries.

Behind the scenes, the European Union climate commissioner, Connie Hedegaard, led the Progressive Ministers Dialogue, also known as the Cartagena Dialogue, in an attempt to create broad support for more ambitious climate action. When her mandate ended in 2014, the Marshallese Minister of Foreign Affairs, Tony de Brum, stepped up to fill the void. He ambitiously raised the stakes and widened membership of the Interministerial Group in preparation for the 2015 Conference of Parties in Paris (COP21).

While no individual could have forged the breakthrough that occurred in Paris, it was de Brum's ambition and vision that made it possible. The Interministerial Group managed to mobilise countries from all negotiating blocks, including the USA and Brazil, to join what became known as the High Ambition Coalition.

'I received an SMS then saying that Brazil would join the HAC and was happy for it to be announced,' recounts Farhana Yamin, who worked with Tony de Brum through this process. The 'high ambition had worked its magic.'[19]

The outcome was an enormous victory that few people dared hope for. It gave the French, who were drafting the agreement, 'the mandate and the credibility to incorporate the most ambitious language possible, including on the 1.5°C global temperature goal, on the need for long-term strategies to reach net-zero emissions by 2050.'[20]

More than half a decade on, the Paris Declaration continues to offer a strong moral and legal compass for climate action, though it has not yet delivered on its promise. Ahead of COP26 in Glasgow, policies put us on a warming trajectory close to 3 degrees (2.1–3.9°C, to be precise). Pledges and targets promise to reduce this to 2.4°C of global warming (within a 1.9–3.0°C range) by the year 2100.[21] Some are more optimistic, arguing that keeping warming below 2 degrees is possible 'if the pledges on the table are implemented in full and on time,' though even they stress that 'commitments made so far, especially for this decade, fall far short of what is required to limit temperature rise to 1.5 °C.'[22] Even so, commitments alone are not enough; they need to translate into decisive action.

Thinking about the future should go well beyond 2100, as researchers have warned that their 'projections show global climate impacts increase significantly after 2100 without rapid mitigation.'[23] This would put the Marshall Islands at severe existential risk.

That risk is pressing enough today. Sea level rises projected to occur this century mean that 40 percent of buildings in Majuro, the Marshallese capital, would be permanently inundated, with fully 96 percent of the city facing frequent flooding. Meanwhile, several of the state's islands could disappear entirely.

'It's always been a dark future, but now that dark future is becoming more clear,' said Kathy Jetñil-Kijiner, a Marshallese poet known for her vocal participation in climate action, about a World Bank report outlining the implications of expected sea level rise in the Pacific. 'One of the islands listed as being 100% underwater, completely covered, is Jaluit, which is actually the island where my family comes from. It's the land that my daughter is named after.'[24]

Beyond the practical and emotional implications of climate change, which are bad enough, the country also faces legal uncertainty. A state cannot exist without an inhabitable territory, so what happens when the country shrinks to such an extent that it can no longer house people? Rights over maritime territory and exclusive economic zones under the 1982 Law of the Sea Convention depend on the position of a state's coastline. But what happens when that coastline permanently shifts, as parts of a state disappear forever?

Territorial fishing grounds, on which many island states rely for food and survival, would become part of the high seas, leaving states with no control over what is now legally theirs to govern.[25]

The Marshallese story is not unique. Many small island states have struggled to convince the international community of the urgent need to take action. In 2009, Mohamed Nasheed, then President of the Maldives, organised a ministerial meeting under water. Equipped with scuba gear, his cabinet signed an SOS message ahead of the climate summit in Copenhagen that year.[26]

It is in response to this existential threat that Pacific Islands, in particular the Marshall Islands, used climate diplomacy to break through the deadlock between 'developed' and 'developing' countries at Copenhagen.

'Diplomacy for us is not an art,' said Tony de Brum in his capacity as Minister of Foreign Affairs for the Marshall Islands, weeks before the COP21 in Paris. 'It is a necessary tool for survival.'[27]

Despite de Brum's diplomatic success in Paris, his homeland remains under enormous threat. This peril drove him to play a crucial role in securing regulation at the IMO.

Prior to the 68th session of the IMO's Marine Environment Protection Committee (MEPC) in 2015, support for shipping decarbonisation was feeble and uncoordinated. The European Union had tried pushing through a target since 2003, but as they had no formal representation, they failed.[28] It did not help that shipping executives simply did not believe the sector would ever be able to decarbonise.[29] Their power over decision-making at the IMO lead to inertia.

When Tony de Brum arrived in London for the meeting, his seat was taken. In practice, IRI (International Registries, Inc.) was calling the shots. This privately run shipping registry, based in the US state of Virginia, was acting in its own interests, not those of the Pacific Island state, which was so urgently threatened by climate change. Those in his place couldn't believe that he represented the Marshall Islands. For them IRI has represented the country for years. As the Marshall Islands is home to the third-largest shipping registry in the world, it wields much power at the IMO. Financial contributions to the UN agency are based on the registered gross tonnage of each member state. While the organisation operates on a one-country-one-vote principle, in practice decisions are made by consensus. Power matters.

IRI now claims on its public website that at the IMO office in London, the Marshall Islands' government delegation is 'augmented by professional staff from IRI offices who contribute to the discussion of issues effecting [*sic*] safety, security, and environmental protection through their shipping knowledge and subject matter expertise.'[30]

At that 2015 meeting, two important things happened. First of all, Tony de Brum claimed his seat at the table. He was, after all, the Minister of Foreign Affairs for the Marshall Islands. Second, his impassioned speech at that 68th MEPC meeting set off a shift in thinking at the IMO.

> Our islands lay [*sic*] just an average of two metres above sea level. Day after day, climate change and the resulting sea-level rise and tropical storms take grip on our homes, on our security and on our livelihoods. My colleagues here from our fellow atoll nation of Tuvalu can tell you what it looks like. And Minister Bule, here all the way from Vanuatu, can tell you how it feels to have 70 percent of your capital city wiped away by a cyclone whose winds were whipped up by the quickly warming Pacific Ocean. Any country here that lives an island existence or that has big populations living along low-lying coastlines can, and will increasingly be the victim of such events.[31]

Despite this urgent call, it would take another three years for the IMO to set a target. At the One Planet Summit, held in Paris two years after the 2015 COP21, several countries signed the Tony de Brum Declaration. Signatory states implored the IMO to adopt an 'initial strategy [that] must set a level of ambition for the sector that is compatible with that of the Paris Agreement, including a peak on emissions in the short-term and then reducing them to neutrality towards the second half of this century.'[32]

The declaration was an ode to the diplomat, who died just months earlier, in August 2017.[33]

Within six months, in April 2018, the IMO did adopt an 'initial strategy.' Unfortunately, while the IMO's target claims to be a 'point on a pathway of CO_2 emissions reduction consistent with the Paris Agreement temperature goals,' its 50 percent minimum target exposes the Marshall Islands to an existential threat from rising sea levels. Indeed, the target is far less ambitious than what the Pacific-led Shipping High Ambition Coalition called for: 'imminent peaking of GHG emissions at 2008 levels, the rapid decline in GHG emissions starting as soon as possible, but no later than 2025, and full decarbonisation (to zero GHG emissions) by 2035.'[34]

The high ambition of the Paris Agreement did not translate into the necessary domestic commitments around the world. And the 'build back better' mantra that dominated the early days of the COVID-19 pandemic has not led to major policy change either.

'Adopting a new set of more ambitious and credible targets that are truly consistent with the goals of the Paris agreement,' implored António Guterres, 'must be an urgent priority for both these bodies [International Civil Aviation Organization and IMO] in the months and years ahead.'[35]

While the IMO's 2018 strategy commits to climate action in line with the Paris Agreement's temperature goals, its minimum requirement is nowhere near ambitious enough to ensure this.

The Marshall Islands (along with the Solomon Islands) continues to demand more ambitious commitments. Ahead of the IMO's 77th Marine Environment Protection Committee in November 2021, they tabled a proposal to adopt a US$100 per tonne of CO_2 equivalent, from 2025 onwards.[36] The carbon price should increase beyond that.

'This would create a strong business case for moving away from fossil fuels towards new zero-carbon fuels,' says Casten Ned Nemra, the Marshallese Minister of Foreign Affairs, 'as well as raising funds to help address the obstacles in this transition, such as making sure new technologies and fuels are ready and available in time and ensuring an equitable transition for developing and climate-vulnerable countries.'[37]

The Pacific Island countries also illustrate a well-known paradox about climate change: the countries least responsible for record-high levels of carbon in the atmosphere are at the greatest risk of the rapidly changing climate. And, beyond their limited responsibility for the unfolding disaster, they are also in a difficult position to take action.

The Marshall Islands and the Maldives both have tiny populations and relatively low carbon footprints per person, of 3.25 and 3.70 tonnes respectively. These levels are nowhere near Qatar (32.42), Kuwait (21.62), or the UAE (20.80), the world's worst emitters. Nor are they anywhere near the levels of Canada (15.50), Australia (15.48), Luxemburg (15.33), or the USA (15.24). Even the self-congratulatory climate champions in the European Union (6.42 average), such as Denmark (5.76) and Germany (8.56) remain much higher. Countries like Sweden (3.54), Croatia (4.06), and Switzerland (4.40) are roughly in the same range. But all of them remain well above, say, Burkina Faso (0.22), the Central African Republic (0.07), or the Democratic Republic of Congo (0.03), the world's lowest emitter.[38]

Most countries, including Small Island Developing States like the Marshall Islands, require both compensation for historical emissions, which disproportionally affect them, and help with funding the transition to clean energy that allows them to build sustainable lifestyles.

While demand for shipping continues to grow, the 'carbon budgets' of individuals are shrinking rapidly. The Berlin-based Hot or Cool Institute has calculated that in order to remain below 1.5°C of warming, our collective emissions will have to come down, per person per year, to *no more* than 3.4t CO_2e by 2030, 1.0t CO_2e by 2050, and eventually 0.4 by 2100.[39] Given the levels of emissions indicated above, there is a lot of work to do.

Growing demand for shipping contrasts starkly with the needed reductions. This means not only that the impacts of immediate consumption should come down drastically in rich countries, but also that the 'invisible' emissions of industrial production and cargo transport need to come down significantly. To meet these targets, changes are required in the way we live, move, and consume.

With fewer than sixty thousand citizens, the Marshall Island's domestic regulations, no matter how ambitious, will have miniscule effects on the planet as a whole. Setting stringent emissions for its own fleet in the absence of global regulations under the IMO would most likely see many ships flagged in the Marshall Islands shift to other registries, which would reduce both its revenues from the registry and the country's influence at the IMO. The Marshallese can only hope that the rest of the world will stop their country from disappearing under rising seas.

Tony de Brum's leadership in climate negotiations showed that small states can be hugely important political actors when it comes to mobilising action. But they will always rely on ambitious and binding legislation in and between large countries.

Speeding up

Every day, I kept on calculating the ETA for Horta using the ECDIS. This made no sense at all when sailing. It's as silly as tracking progress on climate change by your personal consumption habits. You might feel good, but only if you ignore absolutely everything you have zero control over.

Ordinarily, ECDIS' ETA prediction is a useful tool. Ships with engines can set a course, a speed, and stick to it. We could do neither; no one really can when sailing. While we've become a great deal better at predicting winds, we cannot do so with the same precision as running at engine speed.

On June 16, Captain gave us an optimistic weather update. We could expect a smooth and swift passage to Horta. If needed, we could re-provision at Flores, one of the westernmost islands of the Azores. Cook said she was 'proud of us' for handling the period of rationing so well. I'm sure she meant well, even though she then called me 'selectively committed.' That much was probably true. I can work up motivation for just about anything, as long as I want to do it. When I replied that she, too, was selectively committed, she observed that if she was, I'd go hungry. I meant to say that she was selectively committed in her life choices. She had committed to being aboard the *Avontuur*, rather than anywhere else. She was committed to climate action through sailing cargo vessels, rather than any other means. She was committed to the life she'd chosen. But she took it as a personal attack – as *she* had initially meant her sneer at me – and fired back by implying that feeding me was something she'd rather opt out of.

Despite our personality clashes, which were more often friendly than ugly, I greatly admired Cook's commitment. Our collective experience of being stuck at sea inspired her to start writing a manual-cum-recipe book about how to forage at sea, build up supplies, and cook with limited resources and little energy. After returning to shore, she started connecting the sail cargo community, from ship to shore. Her aim is to create a reliable supply network for vessels as they pass through port. (In case you were wondering, throughout the voyage we tried to catch fish, but we did not catch a single one – apart from the flying fish that occasionally landed on deck or in the folds of the reefed sails.)

Rob, the *Hawila*'s cook, is working towards a similar network. Cooks, more than anyone, realise just how fragile life at sea is; how much seafarers rely on supplies from shore.

By June 19, we had 750 nautical miles to go before Horta.

Later that day, we received an update from Cornelius. He wanted us to be in Hamburg by July 22, without delay or excuse. We'd have to unload the cargo before he could get the ship in the dry dock in preparation for an overdue hull inspection.

Joni did not like my noting that there was always a chance that we could not go ashore at Horta. We had, so far, not received any news on whether the mid-Atlantic sailing capital would be open. Going by our previous experiences, I did not want to get my hopes up.

The next day, we celebrated Captain's birthday, by pooling some of our remaining chocolate, which Cook turned into treats to be shared on the aft deck. At that point, Horta was – as I'd feared – still closed for incoming sailors, but Captain heard from Cornelius that the port was about to open up. If borders were to open as expected then, after a health check, we would be able to go ashore. However, I did not know that this would involve a 'brain tickle' through the depths of my nose cavities.

As we were approaching Horta, the mood improved markedly. Every last one of us was looking forward to shore leave. The promise of *terra firma*, cold beer, real showers, fresh food, and alone time beckoned.

Cornelius' 'tight deadline' made the inevitable end of our time together seem a lot more real. After leaving Horta, we would have to maintain a constant speed of four knots, a bit above our actual average speed, in order to make it possible to get to Hamburg in nineteen days, when sailing through the Channel. Any slower than that, we'd have to motor-sail to make up time. First of all, we had to make it to Horta without running out of food.

As Cornelius was too proud to explicitly say we could motor-sail if necessary, and Captain was too proud to ask, it seemed unlikely we'd be able to do this. On June 21, the northern summer equinox, we averaged eight knots, up from just over seven knots earlier. At such speeds, we would not have to motor at all.[40]

At 19:44 that evening, the fun ended. A big wave from aft port went through the open porthole of the aft head. More annoyingly, the water also went down the air hatch leading to the captain's cabin.

He came up from his afternoon nap, understandably in a terrible mood. Our White Watch should have closed the hatch, as the wind had picked up and the sea had gotten rougher. But the splash really did come out of nowhere.

On top of soaking Captain's bunk, the water set off the silent Ship Security Alert System. And it would not turn off until it dried again. Captain had to email Cornelius right away to clarify that we were, in fact, not being attacked by pirates.

Slowing down, scaling down: but will the planet notice?

At €4.20 per kilo, for a one-way trip from Honduras, Belize, or Mexico to Hamburg, our cargo had to be high value. This price was so high because we were sailing a slow and small vessel. But neither slowness nor smallness are valued in today's world. Perhaps they are as an aesthetic, among those who seek meaning in manual craft and the time-consuming skill it demands. For most of the planet's 7.9 billion people, what matters is that there's food on supermarket shelves, medication in pharmacies, and building materials available to construct and maintain shelter.

In their book, *The Ends of the World*, the philosopher Déborah Danowski and anthropologist Eduardo Viveiros de Castro call for 'deceleration' and 'sufficiency,' as do sail cargo companies.[41] In 2022, the notion of 'sufficiency policies' made it into the latest IPCC report, which defines it as 'a set of measures and daily practices that avoid demand for energy, materials, land and water while delivering human wellbeing for all within planetary boundaries.'[42] The political economist and self-declared 'vagabond scholar'

Timothée Parrique argues, after a close analysis of the full report, that 'sufficiency' equals 'degrowth' in all but name.[43]

Why, for some, is the idea of going down to the sea in very slow boats so appealing in a meticulously scheduled just-in-time world? And might it make a difference? The sailing life is sold, by travel agents like Classic Sailing, as an adventure. Escape the tedium of everyday life by tasting some extremes! Without, of course, the abusive captains and frequent deaths that seafarers had to endure when Dana, Newby, or Villiers set sail.

Life at sea is often cold, wet, and miserable. Even when it gets warm, it's hot, humid, and sticky. Shackleton, at least, wasn't lying to his aspiring crew members when calling men (indeed, just men) for a journey to Antarctica, offering 'low wages, bitter cold, long hours of complete darkness.' The odds of a 'safe return' have since improved drastically. Even if he most probably never actually placed the iconic call in any newspaper, it did reflect the culture of sailing at the time.[44] Either way, you're never alone, and work never stops. Hardly decelerating. Even so, the appeal is real. Time is different while sailing. You can learn the craft of sail handling and ship maintenance in a slow and dedicated manner that is barely possible in the frenzied world ashore.

Those *going to sea* are often seeking a radical shift. Those *fleeing land* are often escaping problems, disillusionment, or addiction. Life aboard is simple. Housing, food, schedule, and social life are there. There's no choice but to accept the conditions and go with it. Nearly everything is decided for you and provided collectively. You enter a different temporal space, where the clock sets a rigid and never-ending cadence of watches, but the days, weeks, and months blur. Meanwhile, the craft of sailing, sail making, maintenance, is not only valued, but also part of life. There are no mindless administrative tasks, unless you are the master or an officer, who have to prepare and plan port procedures and cargo operations. For the rest of us, all work was deeply meaningful, existential. Being at sea allows for a true apprenticeship in craft and skill.

We now have time for mindful experiential learning, more so than in 'old days' of sail, when seafarers had to learn skills far more quickly as they were fewer and work was harder.

Working sail today resembles a craft workshop in a world of semi-automated production lines. But that's not an inevitability. The last skippers to operate sailing cargo vessels throughout the twentieth century, including the *Avontuur*'s own Paul Wahlen and the American Captain Lou Kenedy, drove their tiny crews hard.[45] Harder than we ever had to work aboard the *Avontuur*. They ran their businesses in a very competitive environment. They could not afford to hire more hands than they actually needed.

What changed in the meantime is that the remaining 'tall ships' typically run as sail training vessels. That's the model that has proven its worth since the time Alan Villiers bought the *Joseph Conrad* in 1934. Having people pay for the privilege of working aboard a sailing cargo vessel was not new. The stingy Åland-based Finnish shipowner Gustaf Erikson started doing it in the 1920s. But Villiers explored it as a model to run a sailing ship with no cargo.

Today, it's the most common model of running 'tall' sailing ships. Tiny professional crews of experienced sailors welcome 'trainees' aboard, who are often entirely new to sailing. Throughout the voyage, the professionals teach novices the ropes.[46] *Bark Europa*, one of the most striking sail training vessels plying the seas, advertises the experience as a true learning adventure:

> No experience is needed, our crew will learn [*sic*] you everything along the way. Unlike going on a cruise, on Bark Europa you will be going on a hands-on, active sailing adventure. You will be 'on watch' for four hours after which you have eight hours of free time. The permanent crew will give you sail training and you will assist in all sail handling. This involves setting – and taking away the sails by hauling – and easing lines, climbing the rigging to furl or unfurl the sails. You will see that after your weeks on board *EUROPA* you will disembark as full-fledged tall ship sailor.[47]

Rather than using the smallest possible number of paid crew, sail cargo companies operate sail training vessels that transport cargo as well. This has multiple advantages.

The most important is money. Trainees are not simply volunteers. They pay for the privilege. Timbercoast charges *Avontuur* trainees some €50 per day, while Fairtransport asks roughly €70 a day to join the *Tres Hombres* or the *Nordlys*. That's not as much as the more adventure-driven sail training vessels like *Oosterschelde*, at €165 a day or the *Eye of the Wind* at €130 a day. The more expedition-oriented *Tecla* charges about €175 a day, and significantly more for their unique Northwest Passage voyages. The *Bark Europa* is on the cheaper end for long 'blue ocean' crossings at roughly €100 a day, whereas their Antarctica expeditions can cost €400 a day.

Trainee fees of €50 a day generate some eighty thousand euros in revenues from eight trainees over a standard six-month Atlantic round trip. The company could not survive without this income to supplement the €4.20 per kilo they charge for the sixty-five tonnes in the hold.

Does this mean that climate action is the prerogative of the rich, who can pay to volunteer? Mostly, yes. Though some crowdfund their voyage fees or embark on research projects and use research funds to cover the cost.

The biggest value for sail cargo companies is not simply the labour or money that volunteers provide. It is visibility. By sharing our experiences, writing, podcasting, or even just telling friends and family about the experience, we're actively turning the invisible shipping industry into a visible space for climate regulation. No matter how few container ships we see throughout our lives, a sailing cargo vessel in the centre of town raises questions and starts conversations.

For professional seafarers, sail cargo voyages offer respite from charters, where trainees are prone to behaving like tourists, or 'complaining cargo,' keen to get their money's worth. The former voyages attract more engaged crowds, who are switched on and committed to sharing in the purpose of the enterprise. Given the limited opportunities to join sailing cargo vessels, would-be trainees generally outnumber berths. Cornelius, and other shipping company owners, can pick and choose. This allows him to ensure gender balance of crew (even if crews remain overwhelmingly white). At least one trainee from every voyage is pursuing a professional sailing career, Bo'sun told me, creating a larger pool of people trained and committed to sailing cargo.

Even so, the struggle to stay afloat doesn't only affect shipowners. The pay on sail cargo vessels is less than on large commercial ships. Bo'sun, for one, worried about the professional futures of people like him in sail cargo. Would there be enough paid opportunities for him, and others, to make a living from this? Would he be able to purchase his own vessel at some point?

Sailing offers ample time to ponder these questions. But absent connection to shore, there's little opportunity to put ideas that can't be accomplished aboard into practice.

At five knots, a full knot above our voyage average, sailing feels smooth, as if on a new train line. Anything above eight knots felt like the nervous tremble of a high-speed train, as if flying slightly above the sea or the rails.

For comparison, a container ship easily cruises at twenty-four knots. Anything below twenty knots would be 'slow steaming.'[48]

Responsibility for the seas

Resources at sea are always limited. We had no proper showers or warm water, but instead used buckets hosed up from the sea to 'shower' on deck. We flushed the toilets with seawater too, using a bucket. We washed our clothes by hand. Electricity was limited and used up fossil fuels, as the water maker and navigation equipment already used more than the solar panels generated. In comparison with modern container ships, our lives were spartan.

But in comparison to sailing vessels of yesteryear, as well as some other contemporary sailing cargo vessels, we lived luxuriously. The ship is, while at sea, all you have. Any mishap spells trouble. There is hardly a better way of understanding the material needs we have and the limits within which we ought to live. If the Earth is our metaphorical spaceship, as Buckminster Fuller suggested, sailing can help gain insight in the inevitable finitude of planetary resources. The ship, however, can't be run like a planet.

'The ship sails,' as John Mack suggests, 'by the efforts and efficiency of the ship's company, not by [the captain's] authoritative command.'[49]

Captain's great responsibility comes with great power. Seafarers knowingly and willingly surrender their freedoms to the captain. You can join a ship or not. But when you do, Captain's power is yours to endure. Defying that power is mutiny.

The return voyage felt as if it would never end. We were on the move all the time. But we had precious little control over where we were going. It mostly felt like we were stuck in a windy share house with little privacy, not a 180-tonne ship in the middle of the ocean. I was, for the most part, just getting on with everyday life, while doing my part at the perpetual chore-wheel that is the ship's watch system.

For better and for worse. Our arrival was always projected far into an imaginary future. 'One day,' I thought, we'd get there. Despite knowing full well the day we'd leave the ship would come, it did not *feel* like it would ever happen.

Much like a ship requiring a master, we need leadership we can get behind when it comes to running Ship Earth. Today, the idea of a communal future is anathema to the celebration of individual consumer choice. Market fundamentalism has eroded democratic action in favour of the public good, by putting the right of the individual consumer above the collective needs of humanity. Necessary climate action implies giving up some personal freedoms and sovereign rights for the common planetary good. The question is how we can collectively, and democratically, decide on how we get to that future in a way that is decisive, sufficient, and fair.

The climate crisis is urgent, but it doesn't feel quite as urgent as steering a ship in rough weather or shepherding a country through a pandemic. So, our collective willingness to submit consumer choice to democratically set limits appears limited. Decades of relegating climate action to consumers and corporate goodwill has proven ineffective. Moreover, it's increasingly clear that we can't simply innovate our way out of this mess by relying on green growth.

Some may think that the sea is a lawless place, a free-for-all. In many ways this is true, since the distance from land suspends some of the rights and liberties that reign supreme ashore.

The 'regulation of the "global commons" that is the sea beyond the twelve-nautical-mile territorial waters,' argue Liam Campling and Alejandro Colás, 'has largely benefited capitalist firms and economies (rather than, say, small or cooperative fisheries) insofar as the principle of freedom of the seas favoured those maritime nations of the North Atlantic seaboard that had, in previous centuries, created a necklace of overseas colonies connecting up their empires based on commercial capitalism.'[50]

This may be true for the sea, where collective action problems pose threats to the common good. But the ship itself is far from a lawless place. It is bound by strict rules. It is in many ways a benevolent dictatorship.

Ship Earth is an excellent metaphor for the climate crisis that is reshaping the world. Looking after the planet also requires submission to rules, or at least principles, that are bigger than us. And it requires active collaboration.

Unlike ships, the planet on which we live fuels life. Its atmosphere, soil and water, and the precarious balance of chemicals they embody make life possible. But a ship does not have this capacity. Apart from mould, of which the *Avontuur* had plenty, little grows on a ship. At least not in the way that soil and water foster growth.

The planet also regenerates and adapts. James Lovelock's Gaia thesis even suggests that the planet is a living organism, which keeps within a temperature range within which life is possible, and keeps a chemical balance that supports it.[51] A ship needs constant attention and care to fight entropy. The 'world without us,' Alan Weissman argues, would flourish as well – probably better – as with humans around.[52] Ship Earth would thrive. But no maritime ship can battle entropy on its own.

Being aboard allows me to realise the finitude of resources available on the planet and our constant reliance on the food and water it offers us. Even in Jules Verne's nineteenth-century fantasy, *Twenty Thousand Leagues Under the Sea*, Captain Nemo did not manage to survive without coming up for air and restocking from time to time. Just as we could not keep sailing forever, needing to arrive in Horta to restock soon, the planetary free-for-all is also coming to an end.

We can no longer uphold the seventeenth-century colonial *Mare Liberum*: 'Grotius defended free trade with no regard for anything but economic profit,' argues Davor Vidas. While 'massive oceanic trade is central in enabling industrial societies to function,' the search for material consumption that this trade enables is 'out of balance with the need to care for the planet.' We need a paradigm shift. Davor Vidas argues for a shift from our destructive economy to a new normative regime for the Anthropocene that is centred on 'responsibility for the seas.'[53]

Responsibility is not something we can claim, it is something we need to honour. Together. Richard Sennett explores what this could look like in modern societies. He argues that it is through authority, trust, and cooperation that we manage to maintain social relationships. His book is no mere observation. 'We are losing the skills of cooperation needed to make a complex society work,' he warns.[54] Sennett proposes that we can re-learn how to collaborate through 'the craft of making and repairing physical things.'[55] Aboard any ship, that skill is vital.

Whether you're stuck on a ship or on a planet makes little difference. There are limits. Getting through it together is what matters. And I believe that the craft of sailing can help us understand how to tackle climate change, as this requires 'radical collaboration' across differences, as Christiana Figueres suggests.

On the morning of June 23, we finally saw land again. We had not seen any since skirting the Florida coast, on May 27. During our morning watch, we took in the sunrise as we spotted the islands of Flores and Corvo, the westernmost islands of the Azorean archipelago.

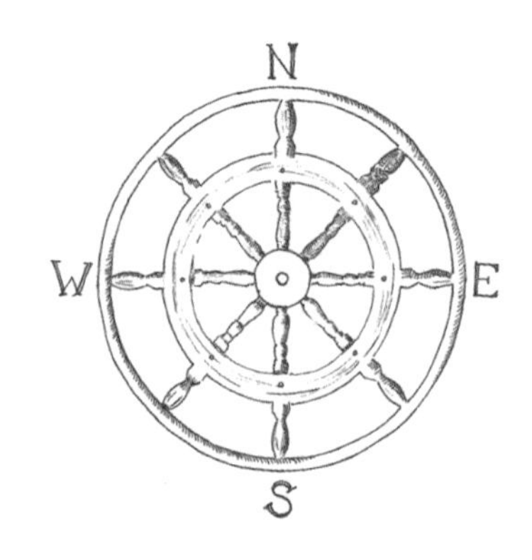

8

Where are we headed?

Horta: first foot ashore

On the morning of June 24, we sailed onto our anchorage at Horta, the mid-Atlantic sailing capital. We dropped anchor outside the breakwater, following a beautifully executed manoeuvre under sail – without ever touching the main engine. The relief of being there, at last, was appreciable. The sight of lush greenery so close was a welcome change to the thousand shades of blue we'd been surrounded by. However close, we were not yet ashore or even in port. The harbourmaster had directed us to the anchorage while we waited to be cleared in. Frustratingly, yachts were allowed to sail into the marina. While we were kept waiting at anchor, their crews went ashore. As a commercial cargo vessel, we were bound by different rules. The regulations in the commercial port required us to enlist an agent. The nearest was in Madeira, over one thousand nautical miles away. This would cost US$1500, at the very least. As we were not calling at Horta to take cargo for a Timbercoast client, this cost would have to be borne by the company. It was, effectively, a fee for a person arranging a parking spot so we could go to the pub and the supermarket. The parking fee itself would follow.

The global roll-out of the 2004 International Ship and Port Facility Security regulations has meant ever-tightening regulations for all commercial ships. These rules are the direct result of the USA's push to increase supply-chain security after the 9/11 attacks. Despite these 'new forms of surveillance of ports, ships, and the people that labour in those places,' ports remain vulnerable nodes in the global supply chain and security infrastructure.[1] On previous visits, the *Avontuur* did not encounter these challenges, so it seems that it took the port of Horta a while to respond to Washington's security concerns.

Stepping ashore, it quickly dawned on me, was not something I wholeheartedly looked forward to. 'Out at sea is the place to be,' was our mantra throughout our voyage in corona times. Being out on the water had spared us from face masks, social distancing, and lockdowns.

After fifty days isolated at sea, getting tested for COVID-19 so we could enjoy shore leave seemed redundant. Without an agent, we weren't even allowed in to do that. Shipping agents take care of the paperwork. They are bureaucratic fixers who ensure ships meet formalities and provisioning while ensuring that port authorities work to the ship's schedule. On our voyage, COVID restrictions meant that agents were our closest contact with the world. They came on board most often, which would not be the case here, if the agent was indeed on Madeira, the other Portuguese island in the Atlantic.

The difficulty and cost of securing port access, combined with the rush Cornelius was in to get us back to Hamburg, made me fear that we'd only go in for provisions and fuel and leave again. Perhaps I was too pessimistic, but I could see it happening.

The delay in going ashore really upset Cook. Her partner was at sea on the *Tres Hombres*, and out of range. Even if we made it ashore, she had no way of speaking to him after those fifty days apart. It felt like he and the rest of the crews on the *Tres Hombres* and the *De Gallant* – two sailing cargo ships – were the only people on the planet who could really relate to our experience. Our lives were radically different from those aboard 'big' cargo ships with their professional crews. And yachties didn't have the same commercial restrictions and cargo routes to follow that we had. Never mind everyone else who'd been cooped up at home.

At anchor that evening, we had dinner together. This was a unique experience, as we never really ate together while at sea. At least, not all fifteen of us. Some were always at the helm or on the lookout while on watch. With land in sight, our conversation turned to the pandemic. There were reasonable and sensible concerns about how this would impact the lives of individuals as well as our collective future, but the tone bordered on that of conspiracy theories, which I'd encounter more often once back ashore.

During dinner, Duarte and Filipe, the water-based crew of *Peter Café Sport*, came out on a RIB with a case of beer, courtesy of the house. *Peter*'s is the Azorean sailor's bar that has served their signature maracuja-infused Gin do Mar to thirsty sailors since 1918. The pair in their white hazmat suits and *Peter*'s beanies, were the first other humans we had seen since leaving Veracruz exactly fifty days earlier. They offered to be our shipping agent. Despite what the harbourmaster told us, there was an agent who was closer than Madeira, which Captain immediately relayed to Cornelius.

The next day, we were told we'd be allowed ashore to get tested the day after. We would get our test results the same evening. No one aboard had any symptoms at all and we'd been at sea for fifty days. So, we were sure to be cleared. But rules are rules, no matter how pointless. That night, we celebrated, still aboard, with our first pandemic-era take-away food, Captain's treat, delivered to the ship by Duarte and Filipe.

Before Bo'sun ferried us ashore on Bob the next day, I recorded a podcast with George Shaw and John Deeks for the Sandringham Yacht Club. I learned as much about life in Melbourne, which was just about to return to 'normal' at that point, as they did about my adventure. Unlike the deserted marina in Melbourne, Horta was filling up with yachts and brimming with activity; many sailors on their eastbound Atlantic crossing, escaping the hurricane season in the Caribbean, called at Horta as well.

Within minutes, we had secured Bob to the pontoon, after which I made my first steps on *terra firma* since leaving Puerto de Mogán on Gran Canaria, one hundred and eighteen days before.

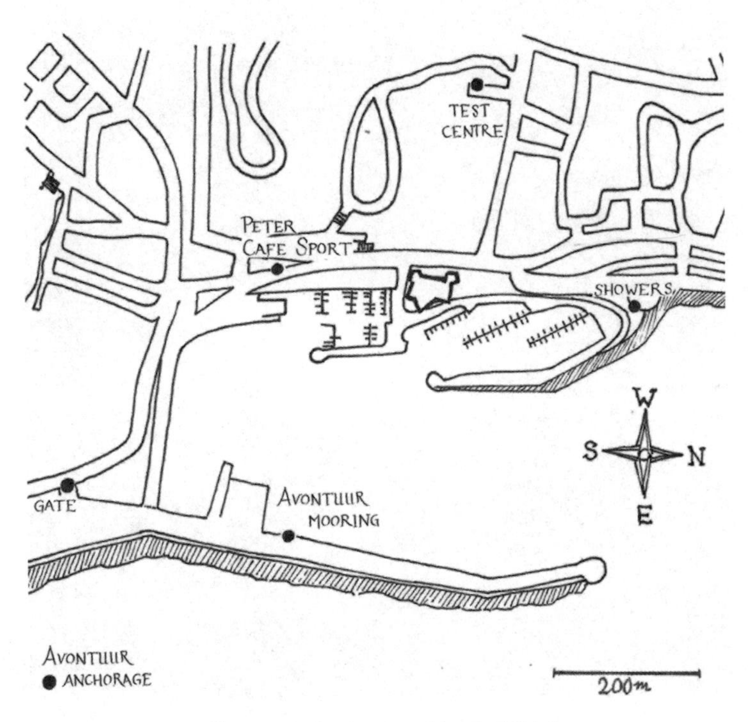

Horta, on the Azorean island of Faial

After a quick stroll past *Peter Café Sport*, to the COVID test centre in a sports complex a few hundred metres up the hill, we queued up for our tests.

'If you're not crying,' one nurse told Joni, 'I'm not far enough up your nose.' I was lucky in getting the less enthusiastic nurse.

The next morning, we were still at anchor waiting to receive our test results. I had a long chat with Cook, whose relationship with Captain was crumbling. Nothing was left of the praise for him she so warmly expressed when leaving Tenerife. Back then, she'd threatened to abandon ship if Captain would leave us. By this point, she was ready to jump ship if Captain stayed on. It's hard to

The well-guarded commercial port compound at Horta, Faial, June 27, 2020

imagine two people less suited to working together, as they could not seem to find a wavelength on which to connect. Most of all, Cook was eager to cook and eat fresh food after our long time at sea. As we were in the dark about when we'd be able to go into port, I suggested she ask Duarte and Filipe to bring us fresh food. She'd then be able to cook a fresh meal that night. At 14:00, before she could act on my suggestion and three full days after dropping anchor, we received permission to come alongside. By 17:00, we were in port, with all our lines secured.

Abandoning protocol, we left the ship at the quay inside the breakwater, went ashore for 'one drink' at *Peter*'s, after which the standing watch had to return. When I returned many hours later, I was the first and only one there.

Next morning, I had my first cafe breakfast, shower, and laundry in four months. To take a shower, we had to walk out of the commercial port, past the marina on the other side, and all the way to the end of the breakwater. There, tucked away in a small corner, was a laundry and showers. For a mere €1.50, one could take a long hot shower. I would have been willing to pay a lot more than that, though it was clean sheets, rather than a mere shower, that made me feel fully human again. A bucket shower on deck always made me feel fresh and clean. But hand washing bedsheets with salt water and rinsing them with a tiny bit of fresh water simply doesn't do the trick.

On June 29, Benji and I drove around the island of Faial, of which Horta is the main town. The First Officer was keen to join us, but Captain did not grant him shore leave that day. He was expecting a delivery of five hundred bottles of gin, carefully packed in boxes of six. They needed to be stowed in the hold. As is common in the merchant marine, the First Officer oversees all cargo operations. Despite this convention, I thought it a bit cruel to keep someone on board for an hour's work after a voyage of one hundred and twenty days, especially since Captain would remain on board all day, along with several other crew.

Without the First Officer, we hiked along the ridge of the Vulcão da Caldeira, which was shrouded in thick fog.

'I have the best hiking shoes money can buy,' Benji would say every hundred metres, 'but they're down there in port.' Within seconds, his sandalled feet were soaked, as were mine minutes later, right through my leather boots.

The Vulcão dos Capelinhos, some ten kilometres west of the Caldeira we'd hiked, is the fascinating site of the island's volcanic eruption in 1957. The eruption extended the surface of the island a few kilometres further west into the Atlantic Ocean. Life on the Azores was hard during the year-long eruption. Harder than usual.

Foggy walk along the ride of the Vulcão da Caldeira, Faial, Açores, June 29, 2020

Benji hiking the ridge of the Vulcão da Caldeira, Faial, Açores, June 29, 2020

Whaling had already become politically and economically unappealing. Arable land was covered in sand and ash. As the island's surface area grew its population shrank. Some four thousand people migrated to Canada and the USA.

On the morning after our trek, Captain insisted that the Red Watch remain aboard until the end of their watch at noon. There was 'urgent' deck work to be done. Even though this work seemed very limited, they had to cancel their plans for time ashore. I later ran into Pinkie, who like everyone else, had grown tired of Captain.

Pinkie had no idea what to tell friends and family back home. It would take some time to get used to the 'brave new world' we faced ashore. Rather than responding to their questions with stories and ideas, she sent them a link to *In the Stillness Between Two Waves of the Sea*, the article Jennifer and I had published in *Ocean Archive* as we left Veracruz.[2]

At 13:00, on June 30, the Second Officer shared a message in our communal WhatsApp group. We were to leave Horta on July 2 at 17:00. We had forty-eight hours left before departure. Until then, we'd remain in our current watches. Our new watches would be posted on the galley door.

Pinkie was not the only one struggling. The First Officer had been unhappy throughout the entire voyage. He was never fully committed to the sail cargo 'intentional community,' and kept feeling like an outsider. He had remained rather private and reserved during the voyage, but during our two months together on the White Watch, he'd opened up a little. Even so, he disliked Captain from day one. Cornelius had even prepared him as they drove to the ship in Tenerife, by saying he hoped the First Officer would warm to, or at least get used to, Captain. That never happened. Despite his lack of sailing skills, First Officer kept up his end of the tacit agreement that he would keep a low profile and never question or challenge Captain. Nor would he make any suggestions or take a proactive role. He simply did what he was told. When I spoke to him that afternoon, he could not hide his disappointment that his compliance had done little to alleviate the tensions between him and Captain.

All at sea

Climate change is in full swing. The global average temperature has increased by 1.2°C since the industrial revolution. And we're headed for 2.7°C (within a 2.0 to 3.6°C range) if we follow current policies.[3] The effects are not a distant possibility for future generations. They are here today. Don't take it from me, or from a literary ornithologist like Jonathan Franzen. Listen to Will Steffen, Johan Rockström, and their colleagues:

> Our analysis suggests that the Earth System may be approaching a planetary threshold that could lock in a continuing rapid pathway toward much hotter conditions – Hothouse Earth. This pathway would be propelled by strong, intrinsic, biogeophysical feedbacks difficult to influence by human actions, a pathway that could not be reversed, steered, or substantially slowed. Where such a threshold might be is uncertain, but it could be only decades ahead at a temperature rise of 2.0°C above preindustrial, and thus, it could be within the range of the Paris Accord temperature targets. The impacts of a Hothouse Earth pathway on human societies would likely be massive, sometimes abrupt, and undoubtedly disruptive.[4]

To put it in more plain, though not necessarily more palatable, language: we do not know at what exact temperature we'll reach more and more irreversible tipping points, but it may well be within 2 degrees. Between the bushfires and the floods, the cyclones and the droughts, climate change is already here. We've been breaking temperature records for as long as I can remember. Those records never last long. We're all at sea with climate change.

Nathaniel Rich explains in *Losing Earth: The Decade We Could Have Stopped Climate Change* that, in 1979, scientists and politicians knew full well the challenge we faced.[5] But they did little. In fact, they deregulated economies and unleashed market forces on the world, rather than reigning them in.[6] To be sure, there have been massive advances in renewable energy but their roll-out has not been nearly fast enough to counter the environmental impacts of unfettered economic growth.

What keeps us afloat is the historic Paris Agreement, which sets a clear objective of keeping temperatures below a 1.5°C threshold. If at all possible. But the action this goal needs is sorely lacking.

If recent trends are anything to go by, record highs will not last for longer than a few years at most before we break them again. And again. These may be the hottest or coldest, wettest or driest months on record, depending where you are. But soon, we're likely to look back on these decades and think of them as marked by comparatively stable conditions.

It's hard to believe that 'net-zero' by 2050, which so many politicians promised us ahead of COP26 in Glasgow, will be enough. It builds on the fiction that carbon dioxide is the only problem. Biodiversity loss and resource depletion may make human life on Earth very difficult, even if global warming can be halted.

When the *Ever Given*, a 400-metre ship carrying tens of thousands of containers, blocked the Suez Canal, the shipping industry made news headlines around the world. Cargo ships are, for the most part, reliable cogs in the globalised just-in-time economy. They ply routes between mines and factories, between producers and consumers, between rubbish bins and dumping grounds. They are the infrastructure that makes global supply chains possible. Shipping is the engine of global trade.

Most of us take invisible infrastructure like running water, electricity, sewers, and ships for granted. As long as infrastructures work, they're out of sight, out of mind; but if you've ever lived in a place where power cuts are common or running water is intermittent, as I have, you'll know that infrastructure rapidly becomes visible when it fails. The obstruction of the Suez Canal and the resulting delays turned shipping into headline news. We are now, more than ever, aware of the near-endless movement of energy, goods, and waste.

The world's poor have long felt the destructive effects of climate change through extreme weather, drought, and environmental degradation. They understand far better than Western city dwellers how fragile infrastructure is.

The shipping industry faces far bigger problems than merely decarbonising. No society reliant on digging up resources and discarding goods at an ever-increasing pace can ever be sustainable. Simply cutting carbon from the process doesn't offer a solution.

Making do with 'local' or 'natural' goods is a chimera. It only makes sense when you blind yourself to how things are made, how much food and shelter your community needs, and how specialised supply chains have become. Shopping 'local' and 'seasonal' is fine. It is even pleasant if you have a supermarket to fall back on. That safety gives us the opportunity to smugly fill one's tote bag with locally grown silver beet at the farmer's market. Because it's a choice.

Let's be honest, if your wealth and comfort depend on the underpayment of other people's labour, as mine certainly does, a few symbolic purchases from a 'local producer' won't save the planet. Nor will it offer absolution when Mephistophélès comes calling on climate judgement day.

Halfway into the voyage, Captain told me that the scale of the *Avontuur* was simply not right. She is too small to be economically viable. The ship should either be smaller (and in a 'lower' category in terms of crew numbers and regulation) or far larger (at least a thousand tonnes) in order to be able to compete. Other companies, like the Dutch EcoClipper and Sail Cargo in Costa Rica, face similar challenges: what is the right size? What is economically viable, while making a difference environmentally? No one really knows. The sail cargo company Grain de Sail in Brittany has opted for a fifty-two-metre ship with a 350-tonne capacity, because it will allow them to enter secondary ports on short notice. This operational consideration matters when operating almost fully under sail, which makes arriving precisely within a narrow time slot – as larger ships in major ports have to – more difficult. Such ships may be economically viable by serving a niche market at a hefty 'green' markup. But how can we bridge the gap between serving sailed coffee to conscientious consumers and decarbonising sixty trillion tonne-miles for every single person on Earth?

The *Avontuur* is a symbolic project. But once larger, purpose-built sailing ships like the *Neoline* or *Oceanbird* go to sea, they'll lose any branding advantages they have. This, as you may have guessed, is the history of shipping repeating itself. Larger and more reliable ships will be able to offer lower rates and shorter transit times, making wind-propelled cargo more appealing for shipowners and cargo owners alike. Or perhaps they'll find a way to generate enough green hydrogen to power the sixty-thousand-strong fleet. And they might have to.

The planet can sustain the lives of all humans that now live on it, but it requires large-scale agriculture, densely packed cities, and advanced medicine, to do so. We can't pick and choose a historical era to 'return' to. We can't 'return' to a random point in the past without accepting that there were far fewer people then. The golden age of sail? That was also the age of opium wars, slavery, and colonialism. There were at most two billion people on the planet.

Historians like Yuval Noah Hariri point out that innovations and improvements allow for population increases, which locks people into new ways of life, from which one cannot simply retreat into the past.[7] Geoffroy Delorme, a present-day incarnation of Henry David Thoreau – or perhaps a voluntary Robinson Crusoe – may have survived as a hunter-gatherer in a forest in Normandy for seven years;[8] it is inconceivable that all sixty-seven million people living in France could follow his example – let alone that they'd want to.

Millions of people are sincerely looking for solutions to the climate crisis, but it's hard to imagine we will find an easy way. At sea, I may not have found the solution. Being out there, with fourteen fellow sailors, I started realising just how large and utterly complex the issue is. Sailing for such a long time, exposed to the elements and moving but at the whim of the winds reminded me of our individual insignificance, our smallness. Our powerlessness in the face of the elements. In the face of such unknowables, many sailors turn to superstition. Even the smallest of things, such as whistling aboard or saying certain words, jinx things; anything that goes against the culture of sailing, so carefully and superstitiously developed over the centuries. But how can we feel so powerless in the face of nature, what the romantics saw as the 'sublime,' while also influencing our environment when simply uttering the word 'rabbit' or whistling?

Enter climate change. It tells us the exact opposite. The aggregate of actions are powerful enough to change the climate by burning fossil fuels. But on our own, we cannot make a difference. Fifteen of us worked around the clock for half a year to ship a paltry sixty-five tonnes of cash crops from former colonies to give do-gooding Germans a guilt-free cup of coffee. So what?

One point five degrees of warming is a given; 2 degrees of warming all but inevitable. Some sailed coffee and chocolate won't change that. Does this mean we should sit back, sip beers, and observe the apocalypse unfold in front of us?

'If we act now'

'If we act now, we can yet put it right,' David Attenborough told the delegates at COP26 in Glasgow. 'We are, after all, the greatest problem-solvers to have ever existed on Earth. We now understand the problem. We now know how to stop the number rising.'[9]

His message echoes what those specialising in climate communication tell us: strike a balance between pointing out the severity and urgency of the situation and the hopeful message that we can still make a difference. It would, after all, be counter-productive to call for urgent measures if all is lost. There is, simply put, no need for climate action, if we're doomed anyway.

'If, working apart, we are a force powerful enough to destabilise our planet,' Attenborough continues, 'surely, working together, we are powerful enough to save it.'[10] We can make it work, if we act *now*.

David Attenborough is far from the only one making this point. If anything, it seems to be the one thing that nearly everyone agrees on. Bill Gates implores us to 'act now.' The Union of Concerned Scientists stresses that 'we've lost precious time, but if we act now – decisively and dramatically – we still have a chance at avoiding climate change's most catastrophic impacts.'[11] The urgency to tackle climate change has increased markedly, and for good reason. But the message remains the same. John Kerry, the US Special Presidential Envoy for Climate, told fellow politicians that Glasgow would be our 'last best hope for the world to get its act together.'[12] Successive Intergovernmental Panel on Climate Change (IPCC) messages make the same point.[13]

I am not so sure. For decades, leaders have been intoning the mantra that immediate action can set the climate straight. Either leaders were wrong about the urgency back then if we still have a chance to turn the tide today; or they are wrong about the feasibility now, as the opportunity has long since passed.

In 1987, Gro Harlem Brundtland's *Our Common Future* report finished with a call to action, stating that 'we are unanimous in our conviction that the security, wellbeing, and very survival of the planet depends on such changes, *now*.'[14] It took a full five years until the landmark 1992 Earth Summit – the United Nations Conference on Environment and Development – took place in Rio de Janeiro.

It took yet another five years, until 1997, for rich countries to agree on carbon emissions reduction targets as part of the Kyoto Protocol. They committed to reduce their emissions against a 1990 baseline. That was the year when the first report of the Intergovernmental Panel on Climate Change came out. Most countries committed to reductions. The USA refused to ratify the Kyoto agreement and Australia negotiated the right to increase emissions.[15]

So, what happened? Between 1750 and 1990, humanity emitted 793 billion tonnes of carbon dioxide. In the thirty years between 1990 and 2020, we emitted the same amount again. Total historic emissions since the onset of industrialisation stand at 1650 billion tonnes. Since I was born in 1986, at 346 ppm, we have collectively emitted as much carbon dioxide as in the 150 years before that.

Waiting for a 'silver bullet' may mean that decarbonisation comes too late. Or it might come with unintended consequences. Indeed, fossil fuels *were* the silver bullet that catapulted a small fraction of humanity into unprecedented material comfort and ease, while creating a capitalist upper class hoarding obscene levels of monetary wealth. It was a Faustian bargain. Mephistophélès is now claiming his due, but we are neither seeking redemption through virtuous striving nor accepting eternal damnation. Our new business-as-usual, captured neatly in the logic of net-zero, is seeking to double down on our collective 'buy now, pay later' approach to the climate.

The Glasgow Climate Pact, the underwhelming result of COP26, stresses the 'urgency of enhancing ambition and action in relation to mitigation, adaptation and finance in this critical decade to address the gaps in the implementation of the goals of the Paris Agreement.'[16]

It is urgent indeed. But despite that urgency, the 'pact' didn't even unequivocally commit to phasing out coal, the dirtiest of all fossil fuels. Forty countries, including Canada, Poland, South Korea, Ukraine, Indonesia, and Vietnam, agreed to stop using coal, but the final text of the conference failed to get every country on board. Instead, governments could only agree to 'accelerating efforts towards the *phase-out* of *unabated* coal power and *inefficient* fossil fuel subsidies.'[17] So, abated coal needs no phasing out, nor do efficient fossil fuel subsidies? Worse, Australia restated its commitment to exporting coal, with its Prime Minister claiming that those working in coal mines will 'continue to be working in that industry for decades to come.'[18]

It's the same old rhetoric: the situation is dire, and urgency is needed, but let's not rush! We need to be 'realistic;' and 'real-world' action needs to be economically 'feasible.'

'Blah blah blah,' Greta Thunberg said.[19]

But what if we looked at climate action differently? 'What if we stopped pretending?' as Jonathan Franzen suggests in a book by that name.[20] He is not suggesting inaction. But he argues we're well past the point at which we can simply halt or reverse the climate crisis. We're in damage control mode now. Even if the 1.5°C and 2°C targets are increasingly out of reach, Andreas Malm argues 'we are still far from the 8°C rise in average temperature' that we'd face when 'burning all of the proven fossil fuel reserves.'[21] Rather than throwing in the towel, we should focus on the gap between those options instead, as they would mean the difference between a 'very dangerous and an unliveable climate.'[22] Simply repeating the same urgent-but-feasible mantra we've heard for decades will only set us up for failure and disappointment. We can no longer invoke the precautionary principle. The damage is already here. It now needs to be limited, halted, and reversed as much as possible.

Many of the countries who joined Tony de Brum's High Ambition Coalition in the lead-up to the Paris Agreement have since 'repeatedly diluted efforts to rein in shipping emissions – with industry representatives in their ears at every step.'[23] At the International Maritime Organization (IMO)'s Maritime Environment Protection Committee, right after COP26, several European countries opposed a 'zero-emissions resolution.'[24] The resolution would, indeed, have increased the IMO's level of ambition. The same shipping companies and governments that called for greater ambition at COP26, blocked greater ambition at the IMO less than a month later. As the sense of urgency seems lost on many, more time is wasted.[25]

I love listening to David Attenborough's soothing voice when he discusses the lives of sloths. But I fear his gentle pressure and calm reassurance is misguided. Along with James Lovelock, I fear that 'the idea that humans are yet intelligent enough to serve as stewards of the Earth is among the most hubristic ever.'[26] We have a lot to learn if we think that simply cutting fossil fuels from our planetary diet will solve the climate crisis.

I won't offer the customary conclusion that 'if we take action now, we can still reverse the damage.' I can't say that in good faith; the evidence is stacked against it. We have now entered damage control mode. We can either 'let it rip,' to use the expression so often used during the pandemic, or we can go all out to limit the damage.

Our voyage aboard the *Avontuur* was a fabulous symbolic action and a life-changing experience, but our environmental impact asymptotically approached zero. Then again, that's the whole principle on which Grotius argued his case for *Mare Liberum*: a ship in the ocean does no harm – though tens of thousands of ships that support consumer capitalism do. Partially by emitting carbon dioxide from propulsion, but mostly by enabling mass consumption built around planned obsolescence.

Sail to the COP?

Greta Thunberg sailed from Plymouth to New York to speak at a climate summit organised by António Guterres, in September 2019.

Detractors criticised her for doing so, as several crew flew across the ocean to sail the *Malizia II* back to Devon. They missed the point. Her voyage was a way of showing how difficult it is to find a means of travel that doesn't cost the earth.[27] She made a principled point, not a claim that sailing the Atlantic was the future of travel.[28]

Thunberg wasn't the only one sailing to a climate conference that year. Thirty-six young climate activists set sail from Amsterdam on October 2. They travelled on the *Regina Maris*, a three-master topsail schooner built in 1908. It would take them seven weeks to reach Rio de Janeiro, after which they'd travel by coach to Santiago de Chile, for COP25. But things worked out rather differently.

Once the *Regina Maris* was well on her way, protests in the Chilean capital led its President, Sebastián Piñera, to pull out of hosting COP25. The thirty-six activists were all of a sudden on their way to an event that had moved. Once they learned the conference was to be held in Madrid, time had run out to change course. Greta Thunberg hitched a ride on *La Vagabonde*, a much faster catamaran on which Australian YouTube celebrities Riley Whitelum and Elayna Carausu then lived.

The anger of these young climate activists is invigorating. It has given seasoned climate scientists and activists a renewed impulse of purpose and meaning. They are angry for good reason. Their parents and grandparents grew up during the great acceleration of consumer capitalism, built on that Faustian pact that offered free labour – in the form of fossil fuels – for immediate wealth and comfort in exchange for a damaged climate in the future. For my own parents and grandparents, this meant greater luxury and ease of living; for the vast majority of the world's population – the Most Affected People and Areas – it meant being lifted from squalor to the modest comfort of secure shelter and sufficient food. In a cynical twist, Faust will be dead by the time Mephistophélès will reclaim his due in full. By that time, Faust's descendants will pay the price for a bargain that previous generations struck.

To make things worse, it is their parents and grandparents – the caregivers supposed to protect them – who have failed so miserably at halting climate change. We now face the challenge of reversing it. Today's parents, simply by living in consumer capitalism, the only life they've ever known, are harming their offspring in ways they can't consume their way out of. Absent urgent action, because the change is so rapid, the children of today's youthful climate activists will face an ever-bleaker future.

Since Greta Thunberg sailed across the Atlantic, concentrations of carbon dioxide in the atmosphere increased from 408.75 ppm in October 2019 to 420.23 in April 2022.[29] The global carbon budget has shrunk by more than a hundred billion tonnes, down from 580 billion tonnes, to stand a 50 percent chance of keeping global warming below 1.5 degrees.[30]

Are we starting to turn the tide? Not at all. After a short-lived slump caused by the pandemic, the International Energy Agency reports that emissions 'rebounded to their highest level in history.'[31]

No one will pretend that COP26 in Glasgow was a success, even if Alok Sharma called it a 'fragile win.'[32] However, shipping is now squarely on the agenda, in a way that it wasn't before the pandemic, in part thanks to supply chain disruptions and the spectacular blockage of the Suez Canal in March 2021. The industry has shown commitment to meet, and exceed, IMO targets. But will their pledges and plans suffice? And will they materialise?

The International Chamber of Shipping (ICS), an industry body that represents the vast majority of shipowners, calls for net zero by 2050 and a carbon levy of US$2 per tonne of fuel to pay for research and development that would aid a technological transition.[33] This proposal is a cynical joke that won't do enough to help shift the transition to clean propulsion technologies (such as wind) and zero-emission fuels, as the paltry amount won't make a dent in the price gap between dirty and clean propulsion mechanisms.

Major cargo owners, like Ikea and Amazon, have since committed to net-zero shipping by 2040, a date sooner than the ICS and IMO. This 'first movers coalition' stresses that liquefied natural gas 'does not meet the criteria' for the emission-free future they envisage.[34] This should be obvious, as LNG is a fossil fuel too. Between carbon emissions and methane slippage, its effects on the climate remain unacceptable – as they may in fact worsen the greenhouse gas effect.[35] The *Ship it Zero* campaign, however, calls the deal 'too weak,' as they call for net zero by 2030 and slam major retailers like Walmart and Target for refusing to join this first mover coalition of the willing.[36]

Meanwhile, the Mærsk Mc-Kinney Møller Center for Zero Carbon Shipping is an industry-driven attempt to chart a pathway for the future of shipping. This includes a levy in the range of

US$50–150 per tonne of carbon, which could help close the gap between fossil fuels and zero-emission fuels. Unlike the European Emissions Trading Scheme, a designated levy for shipping would allow the revenues to remain within the industry, as they argue 'the levy needs just to be large enough to cover the fuel cost difference that the industry faces when switching to alternative fuels.'[37] In other words, as the name of the centre suggests, the only problem that needs solving is carbon.

Fourteen countries, including Denmark, the USA, and the Marshall Islands, have signed a declaration calling for Zero Emission Shipping by 2050. The fifty-five countries behind the Dhaka–Glasgow Declaration echo that call. And the twenty-eight banks that signed up to the Poseidon Principles stated that they will align with a full zero-emission trajectory in 2022.

The World Bank has weighed in by pointing out, mostly in alignment with the Marshall Islands, that the proceeds of a carbon levy could help both the much-needed energy transition and the climate justice agenda by using the enormous levy revenues (tipped to exceed one trillion US dollars by 2050) to help the poorest countries compensate for both higher transport costs and mitigation and adaptation costs for the effects of climate change.[38] It remains unclear whether IMO negotiators will agree on a levy at all. If they do, they might favour a lower levy that covers the energy transition alone, rather than a higher one that would compensate the poorest countries for their climate change losses.

Ambitious as any of these plans are, the question remains whether the shipping industry will be able to act quickly enough. Because the industry has dragged its feet for decades, it has lost precious time. It now has to decarbonise more quickly and radically than many other sectors, precisely because the total carbon budget is running out.

'One barrier' to transforming the shipping industry, the climate scientist Alice Larkin told me, 'is just not really getting that it's about the here and now; really not understanding that.'

At a rate of some fifty gigatonnes a year, the greenhouse gas budget is shrinking fast. Net-zero as a target for 2050 is only realistic if full decarbonisation starts now. Otherwise, the sector will have to move faster and meet the net-zero target sooner.[39]

There is no time to wait until 2030 to start deploying zero emissions vessels at scale. We need, at the very least, 5 percent of the global fleet to run on zero-emission fuels by 2030 to stand a chance at meeting the 1.5 degrees Paris Agreement target.[40] In the meantime, the efficiency of the fleet has to be improved.

'Technologically,' the leading shipping decarbonisation scientist Tristan Smith told me, 'we do know there are solutions like wind assistance, like air lubrication, like optimal steaming speeds and many other types of technologies or operational interventions that can save us a lot more emission and increase the energy efficiency of the fleet.' But mere possibilities won't bring down carbon emissions. The main challenge is that if these existing technologies 'aren't implemented this decade, then by the time we get to the 2030s, we'll have a less efficient fleet, which will need more energy and therefore it'll need more new energy.'

'I'd like to see that plan to create green hydrogen,' Clive Russell of Ocean Rebellion told me. 'How much energy is used in that process is off the scale.' We're yet to find ways to ensure a stable and sufficient supply of new fuels like green hydrogen and ammonia. Russell continued: 'What part of Earth are they going to put all their windmills on, because there's not going to be a lot of space for anyone else to live in, basically?' Bringing down the *energy* intensity of ships is what matters, not just bringing down the *carbon* intensity, because the more zero emissions fuel is needed per tonne-mile, the slower the transition will be.

Where does this leave wind propulsion? As the shipping industry turned to fossil fuels, transport remained largely concentrated along trade winds routes. As a result, there remains 'a good alignment between the windier sea areas and the areas where there is significant shipping activity.'[41]

When Jorne, Andreas, and Arjen founded Fairtransport to set sail aboard the *Tres Hombres* in 2007, no one took wind propulsion seriously. More than a decade of persistence has paid off. Wind propulsion is now a serious contender in the mix of decarbonisation technologies. Not merely as an activist schtick. The International Windship Association, led by Gavin Allwright, has been instrumental in raising the profile of wind-assisted ship propulsion. Danielle Doggett, who is building the *Ceiba* in Costa Rica, has revived and adapted Fairtransport's concept of the *EcoLiner* designed by Dykstra (it now has a lower cargo capacity, but won't rely on any fossil fuels), and plans to build six of them under the name *Veer Voyage* by late 2024.[42]

Since the *Grain de Sail* started her first Atlantic crossing in November 2020, the purpose-built sail cargo vessel has been carrying wine from Brittany to New York, to return with cacao from Latin America.[43] The French region of Brittany is a major hub of activity in the field as several other companies work on a future for wind propulsion, now with support of Loïg Chesnais-Girard, President of the *région Bretagne*.[44] But also the *Oceanbird* by Wallenius Marine and the *Neoline*, both RoRo vessels (short for 'roll-on/roll-off,' meaning that they're designed for cargo that can drive on and off the vessel), are in full development.

A few dozen modestly sized ships won't make much of a dent in emissions, though a rapid roll-out of wind-assist technologies on the existing fleet would. The risk-averse shipping industry may move quickly if these vessels prove effective. Especially with researchers like James Mason of the Tyndall Centre for Climate Change Research in Manchester working on optimising routes for ships using wind-assisted propulsion. His modelling could help improve weather routing software, which is essential for ships that aim to both harness the wind and arrive on time. Better wind conditions may be on a route that is longer. Optimal operation of the Flettner Rotors – cylinders mounted on deck that use wind for propulsion using the Magnus effect – that he studied can also significantly reduce emissions of a voyage, even if it's longer in distance. Albeit making a detour to find favourable winds is risky. The wind does not read the weather forecast, after all. If the weather changes, which is all but inevitable some of the time 'because of the inaccuracies in the forecast data,' Mason pointed out to me in 2021, delays could be 'very severe towards the end of the journey.' As ships would need to speed up to make their booked slot in port or canal, speeding up 'might wipe out all of the savings that you've got,' or worse, the cost of missing allocated slots could prove very high. Better software would help improve the energy gains and emissions reductions from harnessing the wind. Plugging the research gap in this area is both an urgent and highly effective manner to bring industry-wide emissions down quickly.

Will crews know how to operate wind-assisted vessels? Will they know how to rethink their route planning to respond to the – often counter-intuitive – wind predictions? Will they bother? The newly founded wind-assisted ship propulsion course at the Enkhuizen Nautical College is a first attempt to train seafarers to work with these technologies. Without other maritime training centres following suit, wind propulsion may remain niche.

Meanwhile, the debate has shifted. There is now talk of a 'just and equitable transition' in the shipping industry, endorsed by both shipowners and labour unions, which would ensure that the millions of people working in the industry can continue to make a living as the industry transforms. Changing the propulsion technologies and fuels of ships affects maritime workers in ports and on ships more than executives in boardrooms. The need to gain skills and certificates will put increased demands on seafarers and port workers, especially as the industry will likely embrace a mix of fuels and technologies that each come with their own challenges and risks. The Just Transition Maritime Task Force, consisting of the International Chamber of Shipping, the International Transport Workers' Federation, and the United Nations Global Compact, will partner with the International Labour Organization and the

IMO to 'drive decarbonisation of the industry and support millions of seafarers through shipping's green transition.'[45]

The rate at which the industry is embracing change is encouraging. As a result, the IMO should be feeling the pressure by now. Full decarbonisation is far more urgent than their 2018 'initial strategy' suggested.[46] If they don't step up soon, they'll lose their legitimacy and credibility as the global regulator of the shipping industry.[47]

Transport & Environment has argued that because 'parties are to establish "economy-wide" emission reduction targets' and shipping is clearly 'part of our economies,' shipping emissions should be counted in 'nationally determined contributions.'[48] They commissioned an independent legal analysis which concluded that the Paris Agreement requires both shipping and aviation to be included in NDCs.[49] Within a month, the European Commission conceded its plans should 'include measures "for a real transition" in the maritime sector.'[50] But what will that 'real transition' look like? And can we all get behind a common story?

Who needs a compass on an uninhabitable Earth?

If we can't decarbonise shipping, we can't solve the climate crisis. This is not because shipping emissions are so immense. They are not. At roughly 3 percent of global emissions, they're as significant as the aviation industry or a middle-sized country like Germany. If shipping emissions were the only emissions left, however unlikely that is, we'd probably be okay. No, I think our ability to tackle the climate crisis is mirrored by developments in the shipping industry because it is the perfect metaphor for the problem we face.

Cleaning up the shipping industry is a collective-action problem like climate change itself. The governance of the shipping industry through a UN body that is both secretive and heavily influenced by industry reflects just how difficult it is to regulate. The problem is more fundamental. Like climate change itself, shipping mostly happens in the oceanic and atmospheric commons that exist, in our current legal thinking, beyond the sovereign powers of nation states. This has allowed climate negotiators to exclude shipping – and aviation – from global targets that countries commit to implementing. It is possible to allocate responsibility for emissions to countries based on who sells the bunker fuel, who issues the flags, who owns the ship, who operates the ship, or who manages the ships.[51] Though it has proven far more convenient to uphold the legal fiction of the high seas as disconnected from states.

Everyone wants the spoils of global shipping, but no one wants to pay the price. Maybe Garrett Hardin was right after all; maybe the tragedy of the commons is inexorable[52] – though not at the small scale he imagined; Elinor Ostrom has proven him wrong there.[53] It is at a global level where this tragedy unfolds. It is easy to depict the politicians and corporate leaders we rely on to care for the environmental commons as too cynical. It sometimes feels as if they can't imagine anything beyond a tragic wasting of our planetary commons. The myopic economic interests of competing nations trump the common good, but even those who are sincerely committed to addressing the systemic challenge we face in climate change are often impotent for reasons they themselves don't favour. The lack of political will to make difficult decisions in the public interest at the expense of private gains of shareholders might just be the *real* tragedy of the commons.

In many ways, as Andreas Malm cites Theodor Adorno, progress in fighting climate change looks like the 'prevention and avoidance of total catastrophe.'[54] It remains important, however, to distinguish between resisting regression and regressing through inaction, as outsiders to the technical debates, political process, and scientific assessment may understandably find hard to discern.

Shipping symbolises the kind of economy we've built: a gargantuan global patchwork of activities that delivers the goods, at an enormous aggregate environmental cost, both in terms of materials and greenhouse gas emissions. The combination of planned obsolescence and extremely long value chains make the economy, and shipping, far from efficient. Even if the carbon emissions per tonne-mile are small when thinking of just one pair of trainers, simply tinkering and tweaking emissions can't solve the crisis we face. Ever since I set sail on the *Avontuur*, I've grown increasingly convinced that the shipping industry faces a huge challenge. But the potential of such small sailing ships to replace carbon-spewing giants one-for-one is limited at best. I have, however, come to believe that the role of sail cargo lies in something wholly different than merely shipping goods. Rather than simply 'decarbonising' the enormous-and-growing industry, it holds up a mirror. Would we still ship eleven billion tonnes of goods a year if it was as difficult as it was on the *Avontuur*? Probably not. But no single tonne of goods and no single person will make a difference.

We'd need roughly 110 million ships like her to transport all the goods moved by sea each year. That's sixty trillion tonne-miles divided by the 551,330 tonne-miles we transported on the *Avontuur* in 2020. Considering the ship needs a crew of fifteen, working on alternate voyages, that would require more than three billion people, some 40 percent of the world's population, spending half of the year at sea. Year in, year out. Assuming demand does not grow any further. Clearly, that is not realistic.

Economically, politically, and philosophically, if we can't decarbonise shipping, we can't solve the climate crisis. Indeed, as the climate crisis forces us to decarbonise our economies, it also raises questions about the scale of our economic activities. What do we really need? And how much of it? Can we shift to lifestyles in line with a world with just 1.5 degrees of warming? The Berlin-based Hot or Cool Institute think we can and should. In their report, *1.5-Degree Lifestyles: Towards A Fair Consumption Space for All*, the authors argue that 'failing to shift the lifestyles of nearly eight billion human beings means we can never effectively reduce GHG emissions or successfully address our global climate crisis.'[55] They argue we can make the necessary alterations through pursuing systematic change, by lowering the carbon intensity of 'lifestyle options,' or by prioritising changes in citizen behaviour, by lowering amounts of consumption.[56] As I mentioned earlier, this would involve reducing the total carbon footprint per capita to one tonne (tCO_2e) by 2050 and further down to 0.4 by 2100, with 'lifestyle emissions' accounting for 0.7 and 0.3 tCO_2e, respectively.[57]

The shipping industry has two main options.

Either, as most shipping executives do now, it continues to hide behind the fiction that it merely responds to market forces resulting from consumer demand. In this case it cannot play any meaningful role in the systematic change the Hot or Cool report calls for. Shipping bosses would continue to plead innocence in what and how much they transport. The 'cargo manifest' is, after all, sealed from the captain's view.

Alternatively, the industry can embrace the fact that many things cannot be moved around the globe without ships. A great deal of goods would stop being traded if there were no ships to move them. So, as a key bottleneck in global trade, the shipping industry could play a double role in transitioning the planetary economy to a trajectory that aligns with the Paris Agreement temperature targets. If, hypothetically, the industry stops transporting fossil fuels, transport from well to pump would be very constrained. While it would be tricky to put into practice, such an approach would make coal exports almost entirely impossible, as fully 65 percent of coal exports come from two island nations: Indonesia and Australia. Not even previous Australian Prime Minister Scott Morrison is delusional enough to air freight coal out of Australia. (More about this idea in the next chapter.)

When it comes to changing consumption patterns, shipping has one major advantage over individual human beings. There are only some sixty thousand seagoing cargo ships in circulation. Compared to nearly eight billion people, that's a relatively easy number to deal with.

It is difficult to imagine how we could ever do without shipping. Even during the harshest lockdowns throughout the pandemic, ships kept running. Precisely the industry's capacity to keep going when almost everything else ground to a halt symbolises the challenge we face: how can we change everything without changing anything at all?

I started this book by asking where you were during the pandemic; what your lockdown was like. But why use the pandemic as such a marker of time?

Why not ask where you were when the first IPCC report came out in 1990? Or when the Kyoto Protocol was ratified? Or where

Jen reading the compass, while anchored at Puerto Cortés on April 16, 2020

you were during the disastrous Copenhagen COP in 2009? Or, perhaps more optimistically, where you were when world leaders meeting in Paris committed to keeping global warming below 2 degrees, and below 1.5 degrees if at all possible?

As I drafted this in November 2021, the Canadian city of Vancouver was flooded, months after a devastating heatwave and rampant wildfires surrounded the city. British Columbia was far from the only place under water. In the first half of that month alone, major floods occurred in Guatemala, Iraqi Kurdistan, Kalimantan in Indonesia, Jaén in Perú, Badulla in Sri Lanka, Nariño in Colombia, Batu City on the Indonesian island of Java, Bosnia and Herzegovina, Tamil Nadu in India, Egypt (where floodwaters triggered a scorpion infestation), Sardinia in Italy, and the La Paz region of Bolivia. These were the major floods in just those two weeks.

As I revised this in April 2022, Australia had seen months of exceptional rains, while India and Pakistan were plagued by a massive heatwave. This heatwave was so extreme that it nearly resembled the one Kim Stanley Robinson describes in *The Ministry for the Future*'s opening chapter – though that one is set in a fictional future. Not in 2022.

'It is worse, much worse, than you think,' writes David Wallace-Wells at the start of his 2019 book, *The Uninhabitable Earth*. 'The slowness of climate change is a fairy tale.'[58]

The existential threat is more complex than the realisation that a changed climate could make the planet inhospitable, or wholly uninhabitable, for human beings. There is a real risk that we have internalised the idea that we humans are greedy and self-interested. That we are selfish creatures whose future relies on an eighteenth-century Scottish philosopher being right about one very crucial axiom of economic thought: that our self-interest serves humanity as a whole.

This fiction has mutated into an iron law that dominates political thought. 'Neoliberalism's greatest legacy,' Naomi Klein warns us, is that 'the realisation of its bleak vision has isolated us enough from each other that it became possible to convince us that we are not just incapable of self-preservation but fundamentally *not worth saving*.'[59]

Hers is not nearly the bleakest reading.

'Activists,' says Amitav Ghosh in *The Nutmeg's Curse*, 'have long sought to appeal to the conscience of the privileged by emphasising the message that the costs of climate change will largely be borne by the world's poor, mainly Black and brown people.' Perhaps climate change will excite the sense of responsibility for fellow humans that Mia Mottley, the Prime Minister of Barbados, implored us to embrace in her speech at COP26 in Glasgow. She warned that small island nations like hers are in the firing line. Ghosh is not convinced world leaders will heed her call, as the message that poor people will suffer most 'has actually persuaded the privileged to think they need do nothing about climate change because they will be insulated from the worst impacts of global warming by their affluence, and indeed their bodily advantages.'[60]

Some argue the issue is even worse, that we're guilty of what Kari-Mari Norgaard calls implicatory denial: 'the art of professing awareness of climate change while going about one's daily business as though nothing in particular was happening.'[61] Most people are not raving economical maniacs who believe that there really is no alternative to unfettered free markets. It's the normality and banality of implicatory denial that is the vexing problem. Gentle, kind, honourable, moral people convinced that climate change is an issue that urgently needs addressing cause global destruction by living their 'normal' lives. The damage these 'normal' lives do simply don't compute with their inevitable implications. This is why Elizabeth DeSombre asks *Why Good People Do Bad Environmental Things* in her book by that title.[62]

By using colossal amounts of fossil fuels, more and more people have managed to make the most uninhabitable parts of the planet fit for human habitation. From deserts to steppe, billions of people now live comfortably in otherwise inhospitable places. The same fuels have allowed us to live in bigger houses, travel longer commutes, and jet around the world in no time.

Everything is now available all the time, everywhere. Not quite for everyone, not quite everywhere. But few people on the planet are unaffected by the rise of consumer capitalism and the steady rise in standards of living this has enabled.

The pandemic has shown us that we can adopt radical change but not by changing hardly anything at all. Lockdowns stopped travel but freed up money and caused a massive increase in consumer spending and real estate prices in many parts of the world. Left to make our own individual consumer choices, few of us are ready or willing to change our lifestyles. Even if some people are willing to make drastic changes, it is almost impossible to know for certain what 'doing the right thing' looks like in complex societies, as Michael Schur's Netflix series *The Good Place* (2016–2020) points out.

Perhaps ironically, the shipping industry made record profits in 2021.[63]

'Build back better' may win elections. But it's not helping us tackle climate change. Is anyone willing to really cut back on the carbon-intensive habits that we've become so used to? I find it difficult, as I'm keen to reconnect in person with family and friends scattered around the globe.

We can learn from our temporary confinement to our homes, or a ship. We are, as Bruno Latour suggests, confined to planet Earth too. Lockdowns made us rethink how we live in our domestic confines. The climate crisis may help us rethink how we can 'live at home,' on Earth, 'but in a different way.'[64]

As sea levels inch higher, forests disappear, droughts and floods devastate people's lives, and wet-bulb temperatures increase to deadly levels, the planet is indeed becoming uninhabitable. Tinkering with the carbon intensity of gargantuan ships, or even fully cutting fossil fuels, won't change that.

9
Sailing home

The long way home

Sailing is, culturally speaking, a rite of passage. You do not become a sailor by merely stepping aboard. That's only a step towards becoming a sailor. Indeed, no one becomes a sailor until they get back ashore after a first voyage. As our voyage lasted so long, everyone on board – the novices included – clocked more sea miles than many 'hobby' sailors or Sunday-afternoon yachties could hope to get under their belts in a lifetime. Even so, those novices, despite having sailed more than fifteen thousand nautical miles, remained novices until we reached Hamburg.

A sailing voyage starts with a separation from society as a whole; by 'stepping on' and joining the ship, one 'steps out' of life ashore. The ship itself is a 'heterotopia,' a different *space*, but *time* at sea is liminal, in that one's status and role in 'normal life' matters little. The only thing that counts is becoming a fully integrated member of the shipboard *communitas*, which exists only as long as the voyage lasts. A temporary stint ashore does not break this liminality, it merely suspends the arrangements aboard the ship. The moment someone who has been at sea becomes a sailor is when re-joining shore-based life, when re-integrating into society.[1]

Thanks to favourable winds, skill, and coordination, we were able to leave our berth under sail. That is, the rest of the crew were able to do this. I was aboard Bob the dinghy with Bo'sun, giving the *Avontuur* a little push to help her turn one hundred and

Leaving Horta under sail, July 2, 2020

eighty degrees. Sailing a traditionally rigged vessel is a craft. Crews take great pride in being able to smoothly and safely manoeuvre their vessels, especially when fully under sail.

Once the *Avontuur* cleared the breakwater, Bo'sun and I boarded the ship. We sailed off into the Atlantic with the sun setting on the Vulcão do Pico, across from Faial.

Captain had assigned me to the Red Watch, with Martin and Peggy. Once at sea, after a few busy watches, Captain started loosening up again. Our evening watches were very relaxed. After seeing dozens of sunsets while on the White Watch in the tropics between four and eight, we were now enjoying long summer

Leaving Horta, with Bo'sun in Bob the tender, July 2, 2020

Setting the Flying Jib with the Vulcão do Pico in the background, July 2, 2020

days in the North Atlantic with sunsets well after the 20:00 start of our watch. Given the calm seas, I was often alone on deck while at the helm. Captain even regained the sense of humour I'd so enjoyed during the first months at sea. Perhaps I was just nicer to be around.

'Luckily,' Captain said, 'the weather does not read the weather forecast,' as sailing conditions were far better than the daily updates suggested they'd be.

At the first Captain's Reception after leaving Horta, Captain indicated that there would likely be few, if any, pandemic restrictions by the time we'd reach Hamburg.

I wondered what our suspension from life ashore meant for our ritual re-integration. Those on land had lived through their own rite of passage: lockdown. Would we really be able to escape COVID-19 restrictions simply by being at sea during the pandemic? Would life ashore really be 'normal' again when we docked in Hamburg?

Captain proposed, quite helpfully, that when the time should come, we should take a moment to say goodbye to the Atlantic and each other, before getting caught up in the arrival frenzy. Once we arrived in port, it would be difficult to be together alone as a group. As a closing ritual to our voyage, he proposed sharing a last sip of rum in the hold of the ship, once we emptied it of the cargo. All this would take a good while longer; we had barely left Horta. Though he knew, better than most of us, that the chance of all fifteen of us ever gathering again as a group was infinitesimal.

In Hamburg, we'd be surrounded by people keen to hear our stories. Family, friends, journalists, and strangers. Captain suggested trying to anticipate their questions. He seemed wary. There was no way of controlling the narrative of our voyage. We heeded his suggestion. Rather than waiting for someone else to reconstruct our story, we started drafting a joint 'crew statement' that captured our understanding of our adventure. But first, we had to sail two thousand nautical miles from Horta to Hamburg.

During the morning watch on July 6, I told Martin that I'd cleaned the aft head.

'You did not,' he snapped. 'You all just "do" it so you can say you've done it, but you never do it well.' I let his rage blow over, knowing I would never be able to meet his military standards. I clearly lacked *Bundeswehr* discipline. Not that I minded.

The rest of the watch I spent on lookout at the bow. Limited visibility, due to fog, required us to remain far more vigilant about possible obstacles. Captain came forward to tell me I had to look to both port and starboard, as I was perched on the portside of the bow, my sight restricted by the staysail.

With Benji and Jennifer on the aft deck, July 16, 2020

More than simply telling me what I already knew, he came over to make conversation. As so many flights were grounded because of travel restrictions, he said, weather forecasters lacked the data they usually got from civil aircraft as input. He had received an official message from the *Deutsche Wetter Dienst* to this end. The weather forecasts that us 'mission zero' sailors depended on required kerosene-guzzling aircraft. Without accurate weather updates, we were far more likely to get stuck in weather that was too calm or rough to sail. Even so, we were once again expecting a 'fast train' the next day.

'But we don't have a ticket,' Benji joked, 'and the train is full.'

By this point, it was clear that while some aboard could not wait for the voyage to be over, others seemed to hope it would never end. Between the dread of returning to a world crippled by the COVID-19 pandemic, interrupted travel plans, and voluntary disconnection from shore-based life, we all had our reasons. Particularly for those who'd intended disembarking in the Caribbean or Central America before travelling onwards, the prospect of arriving back in Germany wasn't all that appealing. We had covered a lot of distance, but we would soon be back where we had started: in Europe.

Despite the uncertain future awaiting us, the atmosphere was a whole lot more convivial than it had been between Veracruz and Horta. After the start of my evening watch, Benji and I started playing the card game UNO.

'Bingo,' said the First Officer, every time he only had one card left.

I was meant to be working when on duty, but my watch mates were happy to let us have some time for play. Unlike some others on board, we had a healthy understanding that when work called, we answered, but when there was no work, nothing stopped us from socialising.

The characteristic dynamic of sailing crews is that whatever happens on board is either quickly forgotten or sticks until it's stale. Time goes so fast, that we experienced half a lifetime's worth of fun and frustration, sorrow and joy, in just a few months.

Cook thanked me for not giving her a wide berth when she was out of sorts while we were at anchor in Horta, waiting to come alongside, as everyone else had started avoiding her like the plague whenever she was upset.

When we handed over the watch to Pol and his crew at midnight on July 7, Captain told us that, at that point in time, taking the long way home, rounding Ireland and Scotland, was the most favourable option in terms of weather.

That night, while I was helming, I put on Henry Purcell's *The Tempest*. Martin said he was not a fan of such music. 'Since the invention of the microphone,' he reasoned, 'there is simply no need to sing like that anymore.'

Captain came out of the chart room to join us. 'Can't you put on something lively?' he asked. 'This makes me sleepy.'

Martin suggested we could listen to some jazz, which we were all happy with.

After going through the music I had on my phone, I could not find any jazz, so I put on Chico Buarque instead.

'I thought you were going to put on something lively?' came Captain's reply.

'I thought it was a ship rule that whoever is at the helm picks the music,' I replied. To which Captain responded that he'd 'never heard of such a rule.' I calmly explained that this was what we'd been doing since I joined the ship four months prior.

'Not on my watch,' he said, storming off into the chart room.

Peggy looked at me with big eyes that screamed 'You should not challenge Captain's commands!' We passed the rest of that watch in silence. Later on, Captain approached me. He didn't apologise. But neither did I. In his book, an apology was an admission of guilt, of being wrong.

Until that moment, I was happy for the voyage to go on for as long as necessary. From that point onwards, I'd had enough. I started counting the days to Hamburg, a difficult task when you don't know how long it'll take to arrive. The next day, I decided to avoid Captain as much as possible. I stopped asking questions about the route or the weather. I simply followed orders and enjoyed being with others on board the rest of the time.

Around 09:00, we started motoring at about three or four knots, with the engine running at 1000 rpm. We had to get out of the stationary low we'd ended up in. Bo'sun told me that slow motoring like this would use about eight litres of diesel per hour. We kept it running until midnight. At this point, turning on the engine seemed less of a big deal than it had so far. Perhaps it was Cornelius' pressure to make it to Hamburg by July 23. Perhaps it was our own impatience that made it seem like a minor thing to do. Either way, no one seemed to object to motor-sailing half a day.

On July 12, Mia started a new project: counting everything we did and experienced throughout the voyage. While Jennifer and Athena were drafting a beautiful narrative as a proposed joint crew statement, Mia's drive for clarity and numerical detail meant that she felt compelled to count every last thing, down to the birthdays we celebrated on board (ten), air temperatures (min 4.5°C, max 30.5°C), water temperatures (min 0.75°C, max 31.0°C), the number of eggs we ate (eight hundred and twenty-two), the number of lines in the rig (fifty-seven), just to name a few. To her, having a spreadsheet with this information, which she then kindly shared with us, was preferable to having 'subjective memories.' The next day, she broke down in tears. She did not understand our lack of enthusiasm for her plan to compile a record of our 'objective memories.'

On July 14, around 18:30 Athena woke me up from a nap, to ask if I wanted dinner. More importantly, she asked if I knew anything about the Port of Leith on the Firth of Forth, Edinburgh. The latest rumour was that we might make a quick stop there to pick up a barrel of whiskey. As I had lived in Edinburgh, years prior, my questionable knowledge of Scottish waterways was called upon. For a day or two, we had something to get excited about, even if we never actually called at Leith.

The following morning, for the first time on the voyage, several people fell sick. Athena, Jennifer, Joni, Bo'sun, and Pinkie all suffered from the same stomach bug. They thought it could have been from the leftover rice and egg dish they'd eaten. Cook refused to even entertain that idea. Rather than doing so, she served the meal up again at dinner. We finished the meal without complaint, by throwing that part of dinner overboard after she retired to her bunk. This was probably the only time that we did not eat all food, down to the very last scrap.

Cornelius emailed Captain to let us know that the weather might get a bit rough. We'd heard how rough the North Atlantic could get, ever since leaving Veracruz, but we'd been spared such weather throughout the voyage. Barring the occasional squall, we had very few days with high winds. The way Captain spoke about the imminent rough weather made it sound as if we needed it. He had 'promised' us rough weather and we were 'about to get it.'

On July 16, we ceremoniously said goodbye to the Atlantic Ocean, by means of a 'friendship knot,' which contained messages from us. It was a quiet and subdued affair; not quite the cheerful closure I had hoped for.

As we rounded Scotland between Fair Isle and the Orkneys on July 17, we picked up phone connection from shore, along with a whiff of peaty highland air.

Melbourne had just gone into lockdown again to combat its second wave. I had no idea how long it would last. No one had any real way of knowing. Despite the lockdown, I had made up my mind: I would fly home to Australia right away, without spending time in Europe. As much as I wanted to see friends and family, I could not bear the thought of spending a week or two socialising in Belgium and France. I actually looked forward to a fortnight of solitary confinement in a quarantine hotel.

On July 9, we held our last Captain's Reception with brownies. This time around, we had a chocolate surplus. I was not the only one to have some left from our chandler's order in Veracruz. That evening, I asked Captain what his plans were after arrival.

'I don't make plans,' he said. 'I can't make plans like you. I have a ship under my command.'

I did not want to get into a conversation about semantics again. He knew what I was asking. I remained silent, upon which he clarified that Cornelius would take over the ship after we cleaned and disembarked; as Captain he would not leave until the last shipmate had stepped off.

On July 20, as we inched closer to Germany, Captain asked if we should 'slow down' a little, so we could time our passage up the Elbe to coincide with daylight hours. As it turned out, we entered the Elbe in the early morning of July 22, just before the tide would start going in, without deliberately slowing down at all.

I had slept terribly. Reconnecting to the mobile network, coupled with a mix of excitement and anxiety about what life ashore would be like, made it difficult to close my eyes and rest.

As we passed Cuxhaven's Kugelbake (pronounced *bah-kuh* not *bake*), a 'ball beacon' that has guided seafarers into Hamburg's river system for centuries, I remembered sailing past Cuxhaven years earlier on the *Excelsior* when we almost hit a channel marker while navigating the Elbe on our way into the Kiel Canal at night. I gladly accepted Martin's offer to take over the helm from me while we were sailing against the current.

True North

'Mark your head,' Captain would call every hour from the chart room or halfway through the deckhouse companionway. The First Officer would ask the same question with his signature social minimalism; he merely stuck his head out to point at the person helming. That, he must have thought, would do the trick. They were not just curious. Every hour, the captain or officer of the watch needed to fill out the ship's logbook. One of the required entries was the heading, the compass direction in which the ship was steering, at that time.

What Captain was asking was 'Where exactly are we going?' He knew, because he set the course. But he checked in, as a Captain ought, every single hour to make sure we were steering where he wanted us to go. Our collective heading right now, on Ship Earth, is 'net-zero by 2050.' The fundamental issue with this approach leads to what Mike Hulme calls 'climate reductionism,' which leads to 'climate solutionism.'[2] Much like Hulme, I can't see how the planetary crisis can be reduced to a simple metric, whether it's net-zero, 350 ppm, or 1.5 degrees.

In sail cargo, as anywhere in climate action, the tension between big-picture pragmatics and small-scale symbolics remains the struggle of the day. Fair trade has not eliminated exploitative

'Where are we headed?', July 18, 2020

labour practices in coffee or banana plantations. Neither will small sailing vessels decarbonise the shipping industry.

These small vessels have helped to get us talking and reading about maritime cargo transport. We're talking now; but it's high time to decide. What action will we take? I don't believe it's as easy as picking one option over another. Green growth or degrowth? Innovation or tradition? Big or small? Hydrogen or wind?

We're faced with the unintended consequences of modernisation. Jerónimo de Ayanz, Thomas Newcomen, and James Watt did not intend to wreck the planet by developing coal-burning steam engines. Neither did Malcolm Purcell McLean, when easing the loading and unloading of cargo by inventing the shipping container.

History is awash with people trying to improve things. And often they succeed. But with what unintended consequences? Discounting the possibility that any of the options on the table now may cause damage we can't quite conceive of today is precisely what happened last year. And the year before that. And the year before that. This is not to discredit the massive advances we've made in science and medicine. Or the hugely significant ways in which people's livelihoods have improved over the last century. There is now wide agreement about which path we don't want to pursue. That is, we want to avert a climate catastrophe. But there's far less agreement on exactly what we want the future to look like.

Luckily, we have science fiction to help us imagine where we're headed. It certainly helps me think through the options we have. But keeping too tight and narrow a focus on such carefully crafted worlds in the far future may prove counter-productive for two very different reasons.

First of all, the difference between an imagined big-picture future and what we can reasonably accomplish is rather large. This is not because the ideas and aspirations we have are not actually feasible, but because they are abstract. Anna Tsing's anthropological research, which sheds light on the 'friction' between the local and the global, is instructive. She pictures the tension between our actual lives and the abstract ideas we chase with the metaphor of a bridge, which I've referred to earlier. The bridge between our present-day lives and the universal ideas we aspire to is not one we can simply cross, never to look back. It is a 'bridge of universal truths,' which does not simply take us to that abstract ideal world, but to another practical reality.[3] In doing so, crossing that bridge requires us to come 'down to earth,' to use Bruno Latour's words. 'To resist the loss of a common orientation,' he continues, because the modern promise of the 'global' is unattainable and the nostalgic 'local' no longer exists, 'we shall have to come down to earth; we shall have to land somewhere. We shall have to learn how to get our bearings, how to *orient* ourselves.'[4]

The difficulty of finding our place in a world that is undergoing nothing less than a 'metamorphosis' as a result of rapid climate change forces us to rethink the story we tell ourselves.[5] How will we find our way out of this climate crisis? This question brings us to the second reason that focusing solely on a narrowly numerical climate target, however necessary, may not work.

Mike Hulme suggests that these targets overlook the fact that we humans need stories about the past and the future that we can believe in. That future can't be grasped by an abstract numerical target like 350 ppm (or carbon dioxide in the atmosphere), 1.5 degrees (of global surface warming), or net-zero emissions (where creative calculations allow striking a balance between emissions and offsets). Neither can we be certain that any such pathway we choose – if we manage to agree on a choice – would result in a simple binary between success and failure. As it happens, we're skippering between varying degrees of more or less helpful initiatives, whose results are hard to predict.

Even at a more fundamental level, these absolute numerical targets detract from the ways in which we think of the trajectories we're on. Indeed, even the most scientifically minded analysts believe in a story that helps them make sense of their lives. Hulme identifies four main narratives that drive thinking about planet and climate; four ways of imagining the future.[6]

'This is the only way we can pay for this,' is the dominant narrative, or the first story. Growing the economy by further innovation will save the day, 'green growth' proponents argue, conveniently ignoring that it was growth and innovation that landed us in this climate crisis in the first place. They counter this argument by saying that by now, at least we know what the problem is and how we can solve it. These 'ecomodernists' who live by this story believe that theirs is a pragmatic vision that can help us further advance modernity, while lifting all humans out of poverty and keeping the planet liveable.[7] Their utilitarian take rests on a gamble that is very risky. It presumes that we will be able to avoid unintended consequences that could undo any progress we make. It's a risk they're willing to take, because they're convinced it is inherently less risky than any of the alternatives we have.

'This is all the planet can handle,' the second story tells us. It calls for a deep green 'ecological civilisation' built on an idea of restoring a balance with our natural environment, largely by undoing a strict divide between nature and culture. This deontologist approach, built on an active observance of 'do no harm' principles, calls for a return to nature. It is an approach that appeals to many people in theory (who does not want to live a simple life in and with nature?) but appeals to very few people in practice (who wants to live without modern creature comforts?). The underlying idea of this story is that human wants should be brought into line with what the planet can easily support. In arguing for this, it leaves little room for personal liberties to deviate from the eco-centric norm.

'Capitalism can't offer the solutions we need,' is the message of the third story. This story manifests most openly as a 'radical eco-socialist critique of capitalism' as found in the work of Naomi Klein.[8] Extinction Rebellion don't simply make this argument in writing; they take to the streets to demand a structural shift to a new social order, which puts people and the planet before corporate profits. As a narrative of protest, it is powerful and appealing, though the story is stronger on critique than it is on articulating a feasible alternative. The more radical proposals for a Green New Deal align with this perspective.[9] Overall, though, the story tells us more about what is bad than what we can imagine instead.

'This is our only planet, we have to care for it,' is the way the last story frames the world around us. It thinks of the planet as something worth caring for. This story manifests in ancient indigenous civilisations as well as scientific and religious world views.

It sees people as custodians of the land around us. As significant differences exist between the ways in which the beginning of the universe is imagined in this story, so do the stories about its possible ends.[10] This story is a manifestation of virtue ethics, which is a helpful way to guide individuals on how to behave in the world. But it has little to say about the consequences of their actions or the kinds of world that *would* be possible if we wholly embrace this world view.

'To imagine other forms of human existence is exactly the challenge that is posed by the climate crisis,' says Amitav Ghosh. 'For if there is any one thing that global warming has made perfectly clear it is that to think about the world only *as it is* amounts to a formula for collective suicide. We need, rather, to envision *what it might be*.'[11] The tensions between these incommensurable stories hide a bigger truth. No matter how much we believe in our story, we can't really be certain that everyone else is wrong. Each one of us, individually or collectively, is too scared to admit that the story we're peddling is probably not going to offer the solution, whatever it is.

But which of the four stories will it be?

While we ponder our options and rethink the future, I'd like to suggest a few things that could buy us some time. Because we've lost more than enough time already.

Working at the coalface

In *The Human Planet*, the climate scientists Simon Lewis and Mark Maslin convincingly make the case that we do indeed live in the Anthropocene era. They ask: how and when did this era start exactly? Which of the human accelerations was the defining one?[12]

Lewis and Maslin suggest the year 1610 as a starting date for the Anthropocene – a date which conveniently defines the base of the era 'at the beginning of the modern world' that allows us to link 'it to the shift to a capitalist mode of living.'[13]

More fascinating is that 1610 coincides with the expansion of mercantile capitalism, which relied on sailing ships and colonialism. Put another way, the Anthropocene started before the time when steam engines created industrial capitalism – along with its many steamships and empires. Moreover, the arrival of the Anthropocene thus sits well ahead of the emergence of consumer capitalism, which coincides with traces of nuclear bombs in the geological strata of the Earth's crust – the accelerating levels of consumption were made possible by the invention of the shipping container and the global spread of the multi-national corporation.

Neither *mercantile* capitalism, nor *industrial* capitalism, nor *consumer* capitalism would have materialised without the possibilities created by maritime transport across oceans, be it ships propelled by sail, steam, or oil. The geographer Phil Steinberg explores this point in far greater detail in *The Social Construction of the Ocean.* While global capital benefits from the increasingly friction-less ocean space as a conduit for goods, sail cargo companies embrace this friction and celebrate it as something that could slow down trade and with it – they hope – climate change.

By locating the start of the Anthropocene in the early seventeenth century, Lewis and Maslin imply that, bad as our current reliance on fossil fuels is, it was in fact the rise of sail-enabled mercantile capitalism that set off this era in which humans are the defining geologic force. This makes the return to small traditional sailing vessels as a way out of the climate crisis we face a dangerous fantasy.

Lewis and Maslin suggest a way out of the challenge we face by sketching a post-capitalist future characterised by a universal basic income ('or other mechanism(s) that redistribute wealth and resources and also reduce overall resource use levels'), artificial intelligence, solar, wind, wave, and fusion energy. This could put us on a path towards 'environmental and climactic restoration.'[14] This could perhaps halt or reverse our breaking of planetary boundaries.[15]

Unfortunately, we don't have enough planetary time to sip cups of coffee grown by community-supported agriculture and laboriously sailed across the ocean. This kind of thinking and working matters, but we need to simultaneously act to save whatever we can of the remaining carbon budget, and limit the overshoot as much as we can. As shipping historically enabled shifts in society, we might try to replicate the potential of shipping to generate change today. The International Maritime Organization (IMO) is expected to revise its 'initial strategy' in 2023 by increasing its level of ambition and adopting a 'basket of measures' that will offer a feasible transition pathway – though none of this is set in stone.

Ahead of greater ambition at the IMO, I suggest the following five steps. Given the urgency of the issue, I deliberately start with the most unlikely one (halt the transport of coal), to then gradually move to more realistic options, ending with the action that is so obvious, it's shocking it hasn't been adopted yet (slow steaming).

First of all, shipping should act on international agreements by phasing out the transport of coal and the spoils of unsustainable logging. I deliberately say 'shipping,' rather than specifying whom should make this call. Well in line with the Glasgow Climate Pact, this would help to accelerate our 'efforts towards the phase-out of *unabated* coal power.'[16] Shipping company executives might say that they don't decide what gets shipped. That may be so, but without ships, there are no coal exports. I mentioned that the Singapore-based Eastern Pacific Shipping company set an exciting precedent by publicly committing to not carry coal as cargo. The industry is not a mere handmaiden of trade, it is what makes trade possible in the first place. Particularly for fossil fuels.

'In this world, shipmates, sin that pays its way can travel freely, and without a passport,' said Herman Melville, when recounting the story of Jonah and the Whale in *Moby-Dick*, 'whereas Virtue, if a pauper, is stopped at all frontiers.' Maybe it's time to get moralistic about sin that pays its way again.[17]

Imagine, indeed, that the shipping industry would simply stop carrying coal. Not just on one ship, but that no ship would carry coal any longer. That would stop two-thirds of all coal exports because, as mentioned earlier, the biggest two exporters are island nations: Indonesia and Australia.[18] If it seems far-fetched to ban certain cargoes across a globalised industry, it's worth remembering that banning the shipping of slaves was one of the ways in which slavery was curtailed. If US Congress was able to outlaw the importation of slaves through the 1808 Act Prohibiting the Importation of Slaves, it should be possible for countries to unilaterally ban the transport of coal under their flag and seize any vessels trading the commodity in their territorial waters.[19] However unrealistic, such state-sponsored activist tactics would show that their governments are serious about phasing out coal, as promised in Glasgow. This is not nearly as scary as blocking a coal port but it would most likely be far more effective. This first step would immediately cut 10 percent of global shipping demand, on top of the carbon emissions saved from not burning coal. Even if this approach can't stop coal from slipping through the cracks, reducing the global supply by two-thirds would create a massive price increase, making coal-fired plants economically unviable.

Calling on 'shipping,' as an abstract entity that doesn't quite exist, to enact such a ban is perhaps unrealistic, but so is counting on the IMO, which has a mandate to set such a ban, to do so. If anything, the litany of inactions, obstructions, and delays chronicled in this book conveys just how difficult it is to rely on the IMO to take firm and ambitious action. Simply because this proposal might seem unrealistic or even impossible in the current political economy does not make it a bad option. Perhaps the countries that make up the Shipping High Ambition Coalition or the Getting to Zero Coalition could take this proposal forward, as a way to raise the level of ambition beyond what seems possible.

Second, put a carbon price on marine fuel. No more loopholes and exceptions, but a levy that is high enough to make fossil fuels more expensive than other options. This means that the paltry US$2 per tonne of fuel (and thus about a third of that per tonne of carbon) as proposed by the International Chamber of Shipping won't suffice.[20] Several proposals to this end now exist. Ahead of MEPC 76 in June 2021, the Marshall Islands proposed a levy starting at US$100 per tonne of carbon. The Getting to Zero Coalition and the Mærsk Mc-Kinney Møller Center for Zero Carbon Shipping have each published their calculations on how a carbon price would enable an accelerated and complete transition to zero-emission fuels, indicating that support for the measure is increasing.[21]

If at all possible, the carbon levy should act as a buffer to absorb price fluctuations, but only when oil prices fall. When oil prices increase, the total fuel cost increases; but when the oil price drops, the carbon tax absorbs the difference. Once the oil price goes up, it will never come down again. This should give additional security to those investing in zero-emission alternatives. Given the sky-high freight rates today, the industry could easily afford it. A point of contention between the different proponents of a levy is what to do with the funds. The proposals to this end sit between covering the additional fuel cost for first-movers (keeping the funds inside the industry) and using the proceeds of the levy to cover land-side investments in green energy and fuel infrastructure (allowing the funds to be used outside the industry). More radically, the revenues could help fund a 'just and equitable transition' that ensures poor countries, climate vulnerable populations, and those disproportionally affected by both climate change and higher shipping costs are compensated. In the absence of a genuine 'loss and damage' fund for poor countries, this could mean the difference between life and death for many human beings. Even so, the IMO is unlikely to relinquish control over the trillions of dollars this levy could raise in the decades leading up to 2050. Nor is there great interest among IMO negotiators to 'solve the world's problems,' rather than narrowly focus on the decarbonisation of the shipping industry. However, considering that most of the promised hundred billion dollars for the Green Climate Fund never quite made it, this could be an opportunity to fund an energy transition for vulnerable countries.[22]

Third, adopt wind propulsion technologies to the greatest extent and scale possible. This would include retrofitting the existing fleet with the technologies that can help marginal gains across the board. Whether it's a Flettner rotor, a rigid sail, or a kite, anything that helps us bring down emissions immediately will help slow down our inevitable exhaustion of the rapidly shrinking carbon budget. Given the urgency to take climate action, especially in shipping, every percentage counts. As fuel use per ship is enormous, shaving off as little as 10 percent on an annual basis would have a significant impact. The effect of this would be twofold. On the one hand it would help to immediately reduce the carbon intensity of ships, which reduces fossil fuel use and extends the time left to attain full decarbonisation: the most important number is not zero emissions by 2050, but for the remaining carbon budget to stand a chance to keep warming below 1.5 degrees. On the other hand, reducing the fuel intensity of ships by adopting wind propulsion to the greatest extent possible will, in turn, reduce the need for alternative fuels once they become available at scale. Even if fully zero-emission fuels become available, reducing the amount of fuel needed per tonne-mile will help speed up the transition, as it will make a full transition possible more quickly.

Fourth, seriously consider the scale and political economy of the shipping industry. This means developing 'sufficiency policies' in line with the latest Intergovernmental Panel on Climate Change recommendations, which would build on a careful consideration of how to *avoid*, *shift*, and *improve* economic activities.[23] For shipping, such a change would include *avoiding* cargo transport where possible (reducing tonne-miles), *shifting* to wind-propelled vessels and with auxiliary engines using zero-emission fuels, *improving* the energy efficiency of shops and logistics networks more broadly through better hull design, better planning of inventory and routes, and improving the alignment of the most efficient transport speed with supply chain needs. As mentioned previously, Timothée Parrique argues that these 'sufficiency policies' are a thinly veiled call for degrowth in all but name.[24] Thinking beyond growth means decoupling wellbeing from GDP, rather than decoupling GDP from environmental impact.[25] None of this is particularly easy, but these steps allow taking the idea of degrowth seriously, without reducing it to a caricature.

Fifth, if all of this is too wildly controversial, immediately adopt the most easily available operational measures: optimising trade routes and slow steaming. The former would create the additional capacity to compensate for longer travel times. The emissions reductions of the former are up to 38 percent and of the latter could be up to 30 percent.[26] Combined, the gains would obviously be greater.

Meanwhile, we should keep thinking about the future we actually want. That future is very long, both in terms of climate change and artificial intelligence; we are on 'the hinge of history.'[27] What we do now will matter for a very long time into the future, for humans and other creatures alike. The future is very long, indeed. And if we don't solve the climate crisis, we won't need shipping at all.

Captain's stubborn refusal to see 'trying' as an option makes a lot of sense now. We can't simply 'try' to clean up the shipping industry. We have to succeed, no matter what.

However, succeeding isn't as simple as picking a number (350 ppm, 1.5 degrees, or 'net-zero,' which is now *de rigueur*) and sticking to it. We need to know *where* we're going – to mark our collective heading. It remains as crucial as ever to keep an eye firmly on the planetary boundaries we face. But at the same time, we also need to decide *how* we want to go there.

The craft of working together

'We cannot act coherently as a society,' David Korten – a Club of Rome member – stresses, 'without a shared framing story.'[28] And in order to tackle climate change, we need a different story. This is Korten's key message in his 2015 book *Change the Story, Change the Future*.

He argues that today's dominant narrative that explains how the world works is one of 'sacred money and markets.' In this story, time is money:

> Guided by the Sacred Money and Markets story, we have created a global suicide economy designed to make money with no concern for the consequences to life. If our goal is short-term growth of the financial assets of a tiny financial oligarchy, then the system is a brilliant success.[29]

This story created an extractive and profit-driven economy that externalises both planetary and human wellbeing as collateral damage in the pursuit of money. 'The story,' Korten argues, 'is based on bad ethics, bad science, and bad economics.'[30] Despite this deeply entrenched, faulty logic and ethics, the story reigns supreme.

'What the climate needs to avoid collapse is a contraction in humanity's use of resources,' the climate activist Naomi Klein argues. But 'what our economic model demands to avoid collapse is unfettered expansion. Only one of those sets of rules can be changed, and it's not the laws of nature.'[31]

Korten observes that environmentalists increasingly question the narrative that growth and corporate greed, repackaged as 'green innovation,' will deliver the necessary change. 'To become a truly transformative force,' he rightly claims, 'the emerging movement needs a shared public story.'[32] So, what could the alternative story be?

Korten proposes a new meta-narrative, a Sacred Life and Living Earth story, which builds on a wholly different cosmology, where time is *life* and not *money*. This may sound somewhat hippy-dippy for shipping executives, who are used to pushing millions of tonnes of goods across the oceans on gigantic ships. But it is worth entertaining a different narrative and a whole different logic to approach the climate crisis we face. The failures of the dominant Western narrative are too obvious to ignore; Korten argues that we need to shift the focus from wealth accumulation to respect for life as the driving motivation for political and economic action:

> Most every living organism, no matter how small and seemingly unimportant, earns its living contributing to the health, resilience, and creative potential of the whole. All the while, they together maintain a climatic balance and atmosphere suited to the needs of the community's individual members.[33]

His argument is appealing in its simplicity. And he's not the only one making it. In 1972, the scientist James Lovelock used the personification of the Earth in ancient Greece to describe our planet. Gaia, he argued, is 'a complex entity involving the Earth's biosphere, atmosphere, oceans, and soil; the totality constituting feedback or cybernetic system which seeks an optimal physical and chemical environment for life on this planet.'[34] Earth, in other words, is a living creature in her own right, who has inadvertently created a climate in which humans emerged.

The notion that Gaia, a poetic merger of science and mythology, of data and poetry, vindicates indigenous concerns for the planet as 'terra madre,' or 'patcha mama,' giving life, remains contentious. Even so, the idea of Gaia challenges the tension between the 'two cultures' that have caused a rift between the sciences and the humanities, so eloquently discussed by the late C. P. Snow.[35] In a somewhat satirical history of climate change, told from the year 2393, the historians Naomi Oreskes and Erik Conway argue that the academic separation of biophysical and socio-economic drivers of climate change is one of the main obstacles to tackling the climate crisis we're in.[36]

We are now facing up to the challenge of understanding and resisting the 'revenge of Gaia,' in the hope that we understand well enough what we're doing.[37] But do we really know what to do? And do we understand the implications of our action and inaction? Or are we still flying blind?

Sail cargo activists do more than challenge and change the technology used to propel ships. They question the logic, or political economy, of the shipping industry and the globalised economy as a whole. In doing so, they aim to change direction by shifting the narrative that dominates economics. But what exactly is the dominant story and what can we replace it with?

'The problem,' Sam Knights of Extinction Rebellion argues, 'is our complete and utter failure to imagine any meaningful alternative.'[38]

Crews aboard traditional wind-propelled cargo vessels, myself included, do not sign on for the sole purpose of transporting goods. It is the visceral experience of collaborating to accomplish something; it is living with limited resources; and it is living in a dangerous and challenging environment where everything we do comes down to the existential challenge of staying afloat and staying alive.

Jointly voyaging under sail offers a unique blend of climate action that speaks to the time we live in. Sail cargo is also a *positive story*, in that it offers an *alternative*. It does not simply rally against something – even though sail cargo companies certainly take a stand against business-as-usual in the shipping industry. It does not rely on protesting existing practices or even sabotaging them, as has been common in radical climate action movements.

Sail cargo requires active participation through sustained hard work. There are no shortcuts. You can't use an app to increase sail cargo capacity. You need ships, crews, and time. Despite labour-saving technologies, you cannot eliminate any of these.[39]

Despite my self-proclaimed love for loafing and idling, the careful craft of slowly building an alternative speaks to me. Bo'sun approached the craft of sailing in a dialogic manner. He knew how to sail. But he invited us to think differently, experiment, and see if we could find alternative solutions. He, effectively, invited experiential learning; thinking about what we did, why we did it, and if we could do it better. In so doing, we strengthened not only our collective investment in the voyage and the vessel, but also in our social bond.

Through the non-verbal cues we shared while working, we established what Richard Sennett calls the 'social triangle,' by establishing 'authority, trust, and cooperation' through our non-verbal engagements with the ship and each other.[40] Bo'sun subtly forced us to be fully attuned to each other by creating a flow in manoeuvres rather than relying on verbal commands. He oversaw our interactions in silence, observing us, rather than directing our collaboration.

In responding to his evermore subtle cues, we developed a thorough understanding of the ship and the unstable elements that made it move: her crew and the weather. As we could neither control the weather nor alter the ship, our best bet was to respond to each other by attuning our behaviour and by constantly rethinking our role in the social fabric.

At face value, any day on board was the same as any other, dictated by the rigid clock of watches. Like Phil Connors in *Groundhog Day*, we re-lived the same day time and again. We were, like him, at liberty to make minute changes in how we responded to the events that unfolded. By repeating the same interactions, we were able to gently shift our interactions and communal direction.

So, can changing our narrative about the world make a difference? Will *telling different stories* stop a single container ship from transporting enormous quantities of stuff at a huge environmental cost? Probably not, but both technologies we use *and* our narrative of the future have to change.

The *Avontuur* won't decarbonise the entire shipping industry, but we shifted our thinking from avoiding something negative (heavily polluting large-scale shipping) to pursuing something positive (a potential alternative pathway to conducting global trade), however symbolic and small our actions were. We lived through an experiment that aimed to create a real alternative to the path of 'green growth' we're now on.

We were trying to put into practice a *possible* future, which is what activists and social theorists call 'prefigurative politics,' or 'the deliberate experimental implementation of desired future social [and environmental] relations and practices in the here-and-now.'[41]

Perhaps somewhat ironically, sail cargo companies focus so much on climate and consumption, that they forget about labour. Their reliance on professional seafarers willing to work for wages well below industry standards, and volunteers willing to pay for the privilege to work, means that the economic viability of this mode of transport, beyond a few symbolic initiatives, remains a pipe dream.

Indeed, Bo'sun was planning a land-based community project that would make seaborne life easier for crews and shipowners. This would help build skills and craft in shipbuilding, maintenance, agriculture, and living together. His drive stems from the realisation that life at sea is demanding; it is easy to lose oneself in the never-ending work that sail cargo projects require, as is the case in any activist community. Returning to shore after long voyages at sea is made more difficult by the modest income traditional sailing crews get paid – though that's not the only challenge. Many people go to sea to escape land. When on the run from the past, temptation, boredom, or want of opportunity, returning to shore is all the more confronting.

Activists, including those involved in sail cargo initiatives, are often 'unrealistic' in their demands and views. This is strategy, not wishful thinking. Being 'realistic' is precisely what led us to the mess we're in. Indeed, 'realistic' climate action often doubles as business-as-usual inertia.[42] Grassroots activism and engaged entrepreneurialism at the scale of the *Avontuur* may seem antithetical to the massive structural change needed, but it is crucial to gain the political momentum and support for major policy change.

Much like Richard Sennett, I believe that the work of skilful collaboration can inspire us to think beyond one's place of work, beyond a single ship. Indeed, sailing and living together in close quarters elevates cooperation to the status of a craft, which bears lessons for a shared future in an uncertain climate. The challenges we face to repair, remediate, and reconfigure the ship, the community, and the climate are similar.[43]

'Craft,' Bo'sun told me as if he'd just read Sennett's book, 'needs purpose to survive.' That purpose is coming together and repairing our thinking and acting to prepare a workable alternative for the future. That future won't see all our cargo on ships like the *Avontuur.* Nor does it have to. But revalorising craft and realising we're all at sea with climate change is a start.

There remains a tension that is difficult to bridge, as the *Avontuur* is an important symbolic project that resists scaling up due to its size and speed. She is a century-old vessel that was built for an entirely different world. While I look forward to seeing innovative sailing ships become a reality, I am left wondering whether there is a possibility to retain the collaborative potential of traditional sail cargo vessels and the scale of new designs. Newly designed sailing ships present as *technological* projects, where humans can safely remain on the bridge of the ship, operating sails with the press of a button. Their engineers have repeatedly reassured me that these new ships are fully automated and supported by state-of-the-art software that helps tiny crews operate the ships. The design images they share with the world hardly ever feature humans at all. If, in the process of rolling out these vessels, we automate sail to bring down costs and increase efficiency, won't we destroy the social project, which might be central to rethinking the relationship between humans and our environment?

Korten's *Sacred Life and Living Earth* story with which I opened this section is not without its shortcomings. It is, at times, overtly simplistic by suggesting broad agreement between the different schools of environmental thought. In doing so, he reduces conflicting perspectives and objectives to the lowest common denominator that conceals underlying political tensions. But the story must change indeed.

Hamburg

We spent our last night together, anchored in a bend of the Elbe. When aboard a ship, one has little choice in these matters. The port authority tells captains where to berth or moor. It is not like camping, where you can quietly put up a tent at dusk and pack up at dawn.

I was looking forward to being my own boss again. No longer having my time dictated by a watch system, my food choices made by a cook, and my every activity directed by the needs of the ship. In exchange for this freedom, I would be subjected to corona restrictions in Germany and Australia. The lockdown in Melbourne seemed particularly harsh by that point – though the worst was yet to come. Little did I know it would not end until three months later, on October 28.

At midnight, at the end of my anchor watch, Captain told the Second Officer to only wake up the White Watch in the early morning to weigh anchor. That meant I, along with Peggy and Martin, would be able to sleep until 07:00, by which time we should be well underway to the Hafenmuseum in Hamburg.

By the time I woke up, around 06:30, we were on the move, with the *Melpomene* sailing right next to us. At just under twenty metres, the ship looked tiny. Her sails and lines looked like a scale model, despite still having 115 square metres of canvas. A crew of journalists crowded her deck, shooting *Avontuur*'s entry into port.

We motored our way into Hamburg, past ferries, tied-up cruise liners, and container terminals, where we came alongside, onto the quay near the *Deutsches Hafenmuseum*. After an emotional reunion with Cornelius, we started discharging the cargo, aided by dozens of volunteers. It was amazingly strange to see so many new faces all at once.

Shortly after 14:00, after half a day's work as stevedores, when the crane operator was already well into overtime, we hoisted the last net of coffee bags out of the hold onto the quay. It was the same cargo net that served as our 'infinity pool' in the Caribbean. In the empty cargo hold, the fifteen of us spent a last moment together, as Captain had invited us to do weeks prior.

Jen encouraged us to embrace the totality of the experience, both the 'good' and the 'bad,' without romanticising the voyage. I hope this account of our time together does justice to our collective experience. My memory and understanding are necessarily subjective and cannot possibly reflect the views and experiences of all fifteen of us.

When we joined the volunteers for a few drinks at the *Deutsches Hafenmuseum*, Athena read out the 'crew statement' she and Jennifer had drafted. Looking around, no one seemed to understand what we had just gone through, and neither did I.

Arriving was an unsettling experience. Getting back to shore from a sailing voyage always is. But five months is a long time. Life ashore is too easy and smooth in comparison. Too calm and solitary – even in a city bustling with life on a summer day, after opening up post-lockdown.

I started realising how much I'd miss it all. Perhaps not my rank bunk, the food rations, the social rollercoaster, or the pumping and gurgling noises of the head right next to my bunk when sleeping. Being at sea had, by this point, become my 'new normal,' as had the constant movement, both the slow pace through the water and the rapid changes in our communal life.

We celebrated our arrival with a barbecue on the aft deck. The two dozen strangers swarming the deck of the *Avontuur* felt like intruders. They did not belong there. They disrupted the bonds, tensions, and culture we'd created by spending so much time with each other. With only each other.

After a short night, I was hardly able to leave the ship. Freedom beckoned, but it was hardly appealing. What would I do ashore, on my own?

There I was, a few hours later. In my holiday rental flat in the Hanseatic port city of Hamburg. All alone, as if for the first time in my life. At that point, I wanted to disagree with Samuel Johnson's quip that 'a man in a jail,' compared to one at sea, 'has more room, better food, and commonly better company.' Not that I've spent time in jail – barring a couple of nights in a holding cell after the eviction of an Amsterdam squat – but for our experience of being set free in a world that had changed in our absence, we might as well have spent those months behind bars.

Of all the people in the world, I was fortunate to spend five months at sea with these fourteen souls. Cornelius had made T-shirts for us, to commemorate the voyage. On them was emblazoned: 'the world as you know it no longer exists.'

Notes

Annus pandemicus

1 Steven Ujifusa, *Barons of the Sea and Their Race to Build the World's Fastest Clipper Ship* (New York: Simon & Schuster, 2019).
2 Richard Henry Dana Jr, *Two Years Before the Mast* (New York: Harper & Brothers, 1840), 32.
3 M. Anthony Mills and Mark P. Mills, 'The Invention of the War Machine,' *The New Atlantis* 42, no. Spring (2014): 3–23 (3).
4 Alan Villiers, *The Last of the Wind Ships* (New York: Norton, 2000).
5 Cesare Casarino, *Modernity at Sea: Melville, Marx, Conrad in Crisis*, Theory out of Bounds 21 (Minneapolis, MN: University of Minnesota Press, 2002); C. L. R. James, *Mariners, Renegades, & Castaways: The Story of Herman Melville and the World We Live in; the Complete Text*, Reencounters with Colonialism: New Perspectives on the Americas (Hanover, NH: Dartmouth College, 2001).
6 See Nazima Kadir's insightful ethnography of the squatter scene in that neighbourhood, for which she did her fieldwork while I was living there. Nazima Kadir, *The Autonomous Life? Paradoxes of Hierarchy and Authority in the Squatters Movement in Amsterdam*, Contemporary Anarchist Studies (Manchester: Manchester University Press, 2016).
7 Christiaan De Beukelaer, 'The Hundreds of Thousands of Stranded Maritime Workers Are the Invisible Victims of the Pandemic,' *Jacobin*, October 2020.
8 Harry Dempsey, 'Unvaccinated Sailors Risk Deepening Global Supply Chain Crisis,' *Financial Times*, March 22, 2021, www.ft.com/content/72feaabd-5f94-40ad-9073-0178db969b23.
9 UNCTAD, *Review of Maritime Transport* (Geneva: UNCTAD, 2021).

10 Elizabeth R. DeSombre, *Flagging Standards* (Cambridge, MA: MIT Press, 2006).
11 William Barns-Graham, 'Shipping Giant CMA CGN Reports Tenfold Profit Increase with Freight Rates Still Soaring,' The Institute of Export and International Trade, November 2021, www.export.org.uk/news/587994/Shipping-giant-CMA-CGN-reports-tenfold-profit-increase-with-freight-rates-still-soaring.htm; Richard Milne, 'Maersk Warns No End in Sight to Supply Chain Crisis as Profits Soar,' *Financial Times*, November 2, 2021, www.ft.com/content/2ede4d14-dob2-4d85-9daf-420d3f14f9f7.
12 Raphael Baumler, 'Working Time Limits at Sea, a Hundred-Year Construction,' *Marine Policy*, August 2020, 104101, https://doi.org/10.1016/j.marpol.2020.104101.
13 ILO, 'Maritime Labour Convention' (Geneva: International Labour Organization, 2006).
14 Kate Raworth, *Doughnut Economics: Seven Ways to Think like a 21st-Century Economist* (London: Random House Business Books, 2017).

Chapter 1 Departure

1 Daniel Vernick, '3 Billion Animals Harmed by Australia's Fires,' World Wildlife Fund, July 28, 2020, www.worldwildlife.org/stories/3-billion-animals-harmed-by-australia-s-fires; Mark Binskin, Annabelle Bennett, and Andrew Macintosh, 'Royal Commission into National Natural Disaster Arrangements' (Canberra: Royal Commission into National Natural Disaster Arrangements, 2020).
2 Robert G. Ryan, Jeremy D. Silver, and Robyn Schofield, 'Air Quality and Health Impact of 2019–20 Black Summer Megafires and COVID-19 Lockdown in Melbourne and Sydney, Australia,' *Environmental Pollution* 274 (April 2021): 116498, https://doi.org/10.1016/j.envpol.2021.116498.
3 IMO, 'IMO Ship Identification Number Scheme' (IMO, December 6, 2017), 3, A 30/Res.1117, IMODOCS.
4 Hence also the title of Dana's book: *Two Years Before the Mast.*
5 A 'great circle' route is the shortest route between two points on a sphere. On a chart, it looks longer than a straight line because it takes you away from the equator. It is, however, the route ships and aircraft aim for, because it results in the shortest travel time.
6 Charles Heller and Lorenzo Pezzani, *The Left-to-Die Boat*, Documentary (Forensic Architecture, 2012), https://forensic-architecture.org/investigation/the-left-to-die-boat.
7 Daniel Trilling, 'How Rescuing Drowning Migrants Became a Crime,' *The Guardian*, September 22, 2020, sec. News, www.theguardian.com/

news/2020/sep/22/how-rescuing-drowning-migrants-became-a-crime-iuventa-salvini-italy.

8 Raphael Thelen, 'Ships in Mediterranean May Be Ignoring Refugees in Danger,' *Der Spiegel*, November 21, 2018, sec. International, www.spiegel.de/international/europe/ships-in-mediterranean-may-be-ignoring-refugees-in-danger-a-1239495.html; Lorenzo Tondo, '"It's a Day Off": Wiretaps Show Mediterranean Migrants Were Left to Die,' *The Guardian*, April 16, 2021, sec. World news, www.theguardian.com/world/2021/apr/16/wiretaps-migrant-boats-italy-libya-coastguard-mediterranean.

9 Ian Baucom, *Specters of the Atlantic: Finance Capital, Slavery, and the Philosophy of History* (Durham, NC: Duke University Press, 2005).

10 Liam Campling and Alejandro Colás, *Capitalism and the Sea: The Maritime Factor in the Making of the Modern World* (London: Verso, 2021), 6.

11 Ian Baucom, *Specters of the Atlantic.*

12 Kevin Bales, 'Slavery in Its Contemporary Manifestations,' in *Critical Readings on Global Slavery*, ed. Damian Alan Pargas and Felicia Roşu (Leiden: Brill, 2018), 1660–86, https://doi.org/10.1163/9789004346611; Justine Nolan and Martijn Boersma, *Addressing Modern Slavery* (Sydney: University of New South Wales Press, 2019); Michael Odijie, 'Cocoa and Child Slavery in West Africa,' in *Oxford Research Encyclopedia of African History* (Oxford University Press, 2020), https://doi.org/10.1093/acrefore/9780190277734.013.816.

13 Carolyn Merchant, *The Death of Nature: Women, Ecology, and the Scientific Revolution* (San Francisco: HarperOne, 1990), 129.

14 Ariel Salleh, *Ecofeminism as Politics: Nature, Marx and the Postmodern*, Second edition (London: Zed books, 2017); Patrick Curry, *Ecological Ethics: An Introduction* (Cambridge: Polity Press, 2011).

15 Mary Robinson, *Climate Justice: Hope, Resilience, and the Fight for a Sustainable Future* (London: Bloomsbury, 2018), 8.

16 Naomi Klein, *This Changes Everything: Capitalism vs. the Climate* (London: Penguin, 2015), 48.

17 Sindre Bangstad and Torbjørn Tumyr Nilsen, 'Thoughts on the Planetary: An Interview with Achille Mbembe,' New Frame, September 5, 2019, www.newframe.com/thoughts-on-the-planetary-an-interview-with-achille-mbembe/.

18 Ghassan Hage, *Is Racism an Environmental Threat?* (Cambridge; Malden, MA: Polity, 2017), 92.

19 Naomi Klein, *On Fire: The Burning Case for a Green New Deal* (London: Allen Lane, 2019).

20 Andreas Malm and Alf Hornborg, 'The Geology of Mankind? A Critique of the Anthropocene Narrative,' *The Anthropocene Review* 1, no. 1 (April 2014): 62–9, https://doi.org/10.1177/2053019613516291.

21 Norman Myers, 'Environmental Refugees: A Growing Phenomenon of the 21st Century,' *Philosophical Transactions of the Royal Society of London. Series B: Biological Sciences* 357, no. 1420 (April 29, 2002): 609–13, https://doi.org/10.1098/rstb.2001.0953.
22 Klein, *This Changes Everything*, 21.
23 UNHCR, 'UNHCR – Refugee Statistics,' UNHCR, 2022, www.unhcr.org/refugee-statistics/.
24 Roy Scranton in Amitav Ghosh, *The Great Derangement: Climate Change and the Unthinkable* (Chicago, IL: University of Chicago Press, 2017), 130.
25 Andreas Malm, *How to Blow up a Pipeline: Learning to Fight in a World on Fire* (London: Verso Books, 2021).
26 Hage, *Is Racism an Environmental Threat?*, 14–15.
27 Hage, *Is Racism an Environmental Threat?*, 2.
28 United Nations, 'Paris Agreement' (Paris: United Nations, 2015), 2. Emphasis in original.
29 Ghosh, *The Great Derangement*, 158.
30 Amitav Ghosh, *The Nutmeg's Curse* (London: John Murray, 2021).
31 Ghosh, *The Great Derangement*, 148.
32 Ayelet Shachar, *The Birthright Lottery: Citizenship and Global Inequality* (Cambridge, MA: Harvard University Press, 2009).
33 James Lovelock, *Gaia: A New Look at Life on Earth*, Second edition, Oxford Landmark Science (Oxford: Oxford University Press, 2016).
34 Lincoln P. Paine, *The Sea and Civilization: A Maritime History of the World* (London: Atlantic Books, 2015).
35 Brian M. Fagan, *Beyond the Blue Horizon* (London: Bloomsbury Publishing, 2014).
36 Jonathan Raban, *Coasting: A Private Journey* (New York: Vintage Books, 2003), 36.
37 Svante Arrhenius, 'On the Influence of Carbonic Acid in the Air upon the Temperature of the Ground,' *The London, Edinburgh, and Dublin Philosophical Magazine and Journal of Science* 41, no. 251 (April 1896): 237–76, https://doi.org/10.1080/14786449608620846.
38 Climate Action Tracker, 'Climate Summit Momentum: Paris Commitments Improved Warming Estimate to 2.4°C' (Berlin: Climate Action Tracker, 2021), https://climateactiontracker.org/documents/853/CAT_2021-05-04_Briefing_Global-Update_Climate-Summit-Momentum.pdf.
39 Max E. Fletcher, 'The Suez Canal and World Shipping, 1869–1914,' *The Journal of Economic History* 18, no. 4 (December 1958): 556–73 (564), https://doi.org/10.1017/S0022050700107740.
40 United Nations, 'Rio Declaration on Environment and Development' (Rio De Janeiro: United Nations, 1992), Principle 8.
41 https://data.worldbank.org/indicator/EN.ATM.GHGT.KT.CE.

42 Global Monitoring Laboratory, 'Mauna Loa CO2 Monthly Mean Data' (National Oceanic and Atmospheric Administration, 2022), https://gml.noaa.gov/webdata/ccgg/trends/co2/co2_mm_mlo.txt.

43 Johan Rockström et al., 'A Safe Operating Space for Humanity,' *Nature* 461, no. 7263 (September 2009): 472–5, https://doi.org/10.1038/461472a.

44 Will Steffen et al., 'Trajectories of the Earth System in the Anthropocene,' *Proceedings of the National Academy of Sciences* 115, no. 33 (August 14, 2018): 8252–9, https://doi.org/10.1073/pnas.1810141115.

45 IMO, 'Fourth IMO Greenhouse Gas Study' (London: IMO, 2020).

46 IPCC, 'Climate Change Widespread, Rapid, and Intensifying – IPCC,' *IPCC Newsroom*, August 9, 2021, www.ipcc.ch/2021/08/09/ar6-wg1-20210809-pr/.

47 European Commission, 'Delivering the European Green Deal,' Text, European Commission, 2021, https://ec.europa.eu/info/strategy/priorities-2019-2024/european-green-deal/delivering-european-green-deal_en.

48 IMO, 'Initial IMO Strategy on Reduction of GHG Emissions from Ships' (London: IMO, 2018).

49 Transparency International, *Governance at the International Maritime Organization* (Berlin: Transparency International, 2018).

50 Malm, *How to Blow up a Pipeline.*

51 Malm, *How to Blow up a Pipeline.*

52 António Guterres, 'Secretary-General Warns of Climate Emergency, Calling Intergovernmental Panel's Report "a File of Shame" While Saying Leaders "Are Lying", Fuelling Flames,' United Nations, April 4, 2022, www.un.org/press/en/2022/sgsm21228.doc.htm.

53 Will Potter, *Green Is the New Red: An Insider's Account of a Social Movement under Siege* (San Francisco: City Lights Books, 2011).

54 Sam Vincent, *Blood & Guts: Dispatches from the Whale Wars* (Melbourne: Black Inc, 2014).

Chapter 2 What is wrong with the shipping industry?

1 Rose George, *Ninety Percent of Everything: Inside Shipping, the Invisible Industry That Puts Clothes on Your Back, Gas in Your Car, and Food on Your Plate* (New York: Picador, 2014), 92.

2 IMO, 'Fourth IMO Greenhouse Gas Study.'

3 IMO, 'Third IMO Greenhouse Gas Study' (London: IMO, 2015).

4 IMO, 'Fourth IMO Greenhouse Gas Study.'

5 R. Sims et al., 'Transport,' in *Climate Change 2014: Mitigation of Climate Change. Contribution of Working Group III to the Fifth Assessment Report of the IPCC*, ed. O. Edenhofer et al. (Cambridge: Cambridge University Press, 2014), 603.

6 International Transport Forum, 'ITF Transport Outlook 2021' (Paris: OECD International Transport Forum, 2021), 187.
7 Jason Monios and Gordon Wilmsmeier, 'Maritime Governance after COVID-19: How Responses to Market Developments and Environmental Challenges Lead towards Degrowth,' *Maritime Economics & Logistics*, March 14, 2022, 10, https://doi.org/10.1057/s41278-022-00226-w.
8 Joeri Rogelj et al., 'Mitigation Pathways Compatible with 1.5°C in the Context of Sustainable Development' (Geneva: IPCC, 2018), 95.
9 Maria Sharmina et al., 'Decarbonising the Critical Sectors of Aviation, Shipping, Road Freight and Industry to Limit Warming to 1.5–2°C,' *Climate Policy* 21, no. 4 (April 21, 2021): 455–74, https://doi.org/10.1080/14693062.2020.1831430.
10 Climate Action Tracker calculates the impact of international shipping and aviation separately, as if they are countries, because they do not currently count towards national carbon budgets, as other industries do.
11 Climate Action Tracker, 'International Shipping,' 2022, https://climateactiontracker.org/sectors/shipping/.
12 IPCC, 'Climate Change 2022: Mitigation of Climate Change (Summary for Policymakers)' (Intergovernmental Panel on Climate Change, 2022), 42.
13 Taylor Johnson, 'Towards a Zero-Carbon Future,' *Mærsk*, June 26, 2019, www.maersk.com/news/articles/2019/06/26/towards-a-zero-carbon-future.
14 CMA CGM, 'The CMA CGM Group Heads towards Carbon Neutrality by 2050,' June 3, 2020, www.cma-cgm.com/news/3143/the-cma-cgm-group-heads-towards-carbon-neutrality-by-2050.
15 Johnson, 'Towards a Zero-Carbon Future.'
16 Anastassios Adamopoulos, 'Maersk and CMA CGM Push EU for Free Emission Allowances,' *Lloyd's List*, November 30, 2020, https://lloydslist.maritimeintelligence.informa.com/LL1134910/Maersk-and-CMA-CGM-push-EU-for-free-emission-allowances.
17 Mærsk, 'A. P. Møller – Mærsk Accelerates Net Zero Emission Targets to 2040 and Sets Milestone 2030 Targets,' 2022, www.maersk.com/news/articles/2022/01/12/apmm-accelerates-net-zero-emission-targets-to-2040-and-sets-milestone-2030-targets.
18 Faïg Abbasov, 'World's First "Carbon Neutral" Ship Will Rely on Dead-End Fuel,' Transport & Environment, February 23, 2021, www.transportenvironment.org/discover/worlds-first-carbon-neutral-ship-will-rely-dead-end-fuel/.
19 Anastassios Adamopoulos, 'Skou Pitches Minimum $450 per Tonne of Fuel Tax on Ships in the "Medium Term,"' *Lloyd's List*, June 2, 2021, https://lloydslist.maritimeintelligence.informa.com/LL1136996/Skou-

pitches-minimum-$450-per-tonne-of-fuel-tax-on-ships-in-the-medium-term.

20 Marshall Islands and Solomon Islands, 'Proposal for IMO to Establish a Universal Mandatory Greenhouse Gas Levy,' March 10, 2021, MEPC 76/7/12, IMODOCS.

21 Anastassios Adamopoulos, 'Marshall Islands Demands Carbon Pricing Measure from the IMO,' *Lloyd's List*, November 16, 2020, https://lloydslist.maritimeintelligence.informa.com/LL1134705/Marshall-Islands-demands-carbon-pricing-measure-from-the-IMO.

22 Milne, 'Maersk Warns No End in Sight to Supply Chain Crisis as Profits Soar.'

23 Simon Bullock, James Mason, and Alice Larkin, 'The Urgent Case for Stronger Climate Targets for International Shipping,' *Climate Policy* 22, no. 3 (2022): 301–9, https://doi.org/10.1080/14693062.2021.1991876.

24 IMO, 'Initial IMO Strategy on Reduction of GHG Emissions from Ships,' 5.

25 Bullock, Mason, and Larkin, 'The Urgent Case for Stronger Climate Targets for International Shipping.'

26 Bullock, Mason, and Larkin, 'The Urgent Case for Stronger Climate Targets for International Shipping.'

27 Robert Greenhalgh Albion, 'The Timber Problem of the Royal Navy 1652–1862,' *The Mariner's Mirror* 38, no. 1 (January 1952): 4–22, https://doi.org/10.1080/00253359.1952.10658102.

28 Robert Greenhalgh Albion, *Forests and Sea Power: The Timber Problem of the Royal Navy* (Cambridge, MA: Harvard University Press, 1926), vii–viii.

29 Aleksandr Ėtkind, *Nature's Evil: A Cultural History of Natural Resources*, trans. Sara Jolly, New Russian Thought (Cambridge: Polity Press, 2021), 17.

30 Jed O. Kaplan, Kristen M. Krumhardt, and Niklaus Zimmermann, 'The Prehistoric and Preindustrial Deforestation of Europe,' *Quaternary Science Reviews* 28, no. 27–8 (December 2009): 3016–34, https://doi.org/10.1016/j.quascirev.2009.09.028.

31 Ėtkind, *Nature's Evil.*

32 Albion, 'The Timber Problem of the Royal Navy 1652–1862,' 21.

33 Nicholas Stern, *The Economics of Climate Change: The Stern Review* (Cambridge: Cambridge University Press, 2006), xviii.

34 Sailcargo Inc., www.sailcargo.org/.

35 The actual rate of sequestration is higher, as every tonne of dry timber contains about 1.8 tonnes of carbon.

36 Hawila Project, www.hawilaproject.org/hawila-history/.

37 http://classics.mit.edu/Plutarch/theseus.html.

38 Hawila Project, 'The Fascinating Story of Hawila's Wood for the Refit 2020/2021,' www.hawilaproject.org/the-fascinating-story-of-hawilas-wood-for-the-refit-2020-2021/.
39 Paul Jepson and Cain Blythe, *Rewilding: The Radical New Science of Ecological Recovery*, Hot Science (London: Icon Books, 2020); George Monbiot, *Feral: Rewilding the Land, the Sea, and Human Life* (London: Penguin Books, 2014); Isabella Tree, *Wilding: The Return of Nature to a British Farm* (London: Picador, 2019).
40 Sekula's mention of the 'Super Panamax' ship's capacity predates the opening of the new 'Neo Panamax' locks in 2009. In his notes, 'super' thus refers to the size of a ship exceeding the pre-2009 Panamax classification.
41 Sally Stein, '"Back to the Drawing Board": Maritime Themes and Discursive Crosscurrents in the Notebooks of Allan Sekula,' in *OKEANOS*, by Allan Sekula, ed. Daniela Zyman and Cory Scozzari (Berlin: Sternberg Press, 2017), 60–89.
42 Fred Pearce, 'How 16 Ships Create as Much Pollution as All the Cars in the World,' *Daily Mail*, November 11, 2009, www.dailymail.co.uk/sciencetech/article-1229857/How-16-ships-create-pollution-cars-world.html.
43 James J. Corbett et al., 'Mortality from Ship Emissions: A Global Assessment,' *Environmental Science & Technology* 41, no. 24 (December 1, 2007): 8512–18, https://doi.org/10.1021/es071686z.
44 IMO, 'Fourth IMO Greenhouse Gas Study,' 8.
45 Olli-Pekka Hilmola, *Sulphur Cap in Maritime Supply Chains* (Basingstoke: Palgrave Pivot, 2019).
46 Hilmola, *Sulphur Cap in Maritime Supply Chains*.
47 Martin Stopford, *Maritime Economics*, Fourth edition (Hoboken: Taylor & Francis, 2013), 24.
48 UNCTAD, *Review of Maritime Transport* (Geneva: UNCTAD, 2020).
49 Frank Trentmann, *Empire of Things: How We Became a World of Consumers, from the Fifteenth Century to the Twenty-First* (New York; London; Toronto: Harper Perennial, 2017), 688.
50 Trentmann, *Empire of Things*, 690.
51 Lewis Akenji et al., *1.5-Degree Lifestyles: Towards A Fair Consumption Space for All* (Berlin: Hot or Cool Institute, 2021), https://hotorcool.org/wp-content/uploads/2021/10/Hot_or_Cool_1_5_lifestyles_FULL_REPORT_AND_ANNEX_B.pdf.
52 Mar Viana et al., 'Impact of Maritime Transport Emissions on Coastal Air Quality in Europe,' *Atmospheric Environment* 90 (June 2014): 96–105, https://doi.org/10.1016/j.atmosenv.2014.03.046.
53 James Ellsmoor, 'Cruise Ship Pollution is Causing Serious Health and Environmental Problems,' Forbes, April 26, 2019, www.forbes.com/sites/

jamesellsmoor/2019/04/26/cruise-ship-pollution-is-causing-serious-health-and-environmental-problems/; Susan Scutti, 'The Air Quality on Cruise Ships is so Bad, It Could Harm Your Health, Undercover Report Says,' CNN, January 26, 2019, www.cnn.com/2019/01/24/health/cruise-ship-air-quality-report/index.html.

54 Emy Demkes, 'The More Patagonia Rejects Consumerism, the More the Brand Sells,' The Correspondent, April 28, 2020, https://thecorrespondent.com/424/the-more-patagonia-rejects-consumerism-the-more-the-brand-sells.

55 Patagonia, 'Don't Buy This Jacket, Black Friday and the New York Times,' November 25, 2011, https://eu.patagonia.com/gb/en/stories/dont-buy-this-jacket-black-friday-and-the-new-york-times/story-18615.html.

56 Sharon J. Hepburn, 'In Patagonia (Clothing): A Complicated Greenness,' *Fashion Theory* 17, no. 5 (November 2013): 639, https://doi.org/10.2752/175174113X13718320331035.

57 Hepburn, 'In Patagonia (Clothing): A Complicated Greenness.'

58 Allan Sekula and Noël Burch, *The Forgotten Space*, 2010, www.theforgottenspace.net/.

59 Bill McKibben, 'It's Not Science Fiction,' *The New York Review*, December 17, 2020, www.nybooks.com/articles/2020/12/17/kim-stanley-robinson-not-science-fiction/.

60 António Guterres, 'Our Common Agenda' (New York: United Nations, 2021).

61 Thomas Hale, 'Our Common Agenda – Governing the Future?' *Global Policy*, September 10, 2021, www.globalpolicyjournal.com/blog/10/09/2021/our-common-agenda-governing-future.

62 ECDIS, or the Electronic Chart Display and Information System, is a digital chart system for ships, which is acceptable as a replacement for paper charts. Keeping track of one's route, whether on paper charts or a digital system, is a legal requirement on ships. ECDIS is, in other words, effectively a fancy GPS or satnav system for ships.

63 Michel Foucault, 'Des Espaces Autres,' trans. Jay Miskowiec, *Architecture / Mouvement / Continuité*, no. 5, (October 1984): 9.

64 John Mack, *The Sea: A Cultural History* (London: Reaktion Books, 2011), 141.

Chapter 3 Crossing the Atlantic

1 Hannah Arendt, *The Human Condition*, Second edition (Chicago IL: University of Chicago Press, 1998).

2 Lord Admiral Collingwood, cited in Mack, *The Sea*, 158.

3 Dana, *Two Years Before the Mast*, 11–12.

4 Andreas Malm, *Fossil Capital: The Rise of Steam-Power and the Roots of Global Warming* (London: Verso, 2016).
5 Joshua Krook, 'Whatever Happened to the 15-Hour Workweek?' *The Conversation,* October 9, 2017, http://theconversation.com/whatever-happened-to-the-15-hour-workweek-84781.
6 Tim Jackson, *Prosperity without Growth: Foundations for the Economy of Tomorrow,* Second edition (London: Routledge, 2017).
7 Paul Lafargue, *The Right To Be Lazy* (Chicago IL: Charles Kerr and Co, 1883), Appendix.
8 Timothée Parrique, 'The Political Economy of Degrowth' (PhD, Clermont-Ferrand & Stockholm, Université Clermont Auvergne & Stockholm University, 2019), https://tel.archives-ouvertes.fr/tel-02499463; Simon Mair, Angela Druckman, and Tim Jackson, 'A Tale of Two Utopias: Work in a Post-Growth World,' *Ecological Economics* 173 (July 2020): 106653, https://doi.org/10.1016/j.ecolecon.2020.106653.
9 David Graeber, *Bullshit Jobs* (New York: Simon & Schuster, 2018).
10 Christiaan De Beukelaer, 'Tack to the Future: Is Wind Propulsion an Ecomodernist or Degrowth Way to Decarbonise Maritime Cargo Transport?' *Climate Policy* 22, no. 3 (2022): 310–19, https://doi.org/10.1080/14693062.2021.1989362.
11 Simon Bullock et al., 'Shipping and the Paris Climate Agreement: A Focus on Committed Emissions,' *BMC Energy* 2, no. 1 (December 2020): 5, https://doi.org/10.1186/s42500-020-00015-2.
12 Wind-Assisted Ship Propulsion, 'New Wind Propulsion Technology: A Literature Review of Recent Adoptions' (Wind Assisted Ship Propulsion, 2020).
13 Paul Berrill, 'Sail Cargo Reveals Plans for Six 100-TEU Emission-Free Ships,' TradeWinds, November 1, 2021, www.tradewindsnews.com/technology/sailcargo-reveals-plans-for-six-100-teu-emission-free-ships/2-1-1091765.
14 Hellenic Shipping News, 'EcoClipper Purchases Its First Sail Cargo Vessel,' January 21, 2022, www.hellenicshippingnews.com/ecoclipper-purchases-its-first-sail-cargo-vessel/.
15 Theodore Zeldin, *Conversation* (Mahwah, NJ: HiddenSpring, 2000).
16 Ghosh, *The Great Derangement,* 154.
17 Chloé Farand, 'Anger as UN Body Approves Deal That Allows Ship Emissions to Rise to 2030,' Climate Home News, November 17, 2020, www.climatechangenews.com/2020/11/17/anger-un-body-approves-deal-allows-ship-emissions-rise-2030/ (no page).
18 Fiona Harvey, 'Campaigners Criticise Global Deal on Carbon Emissions from Shipping,' *The Guardian,* October 23, 2020, sec. Environment, www.theguardian.com/environment/2020/oct/23/green-groups-condemn-proposals-to-cut-shipping-emissions.

19 Rebecca Ratcliffe and Michael Standaert, 'China Coronavirus: Mayor of Wuhan Admits Mistakes,' *The Guardian*, January 27, 2020, sec. World news, www.theguardian.com/science/2020/jan/27/china-coronavirus-who-to-hold-special-meeting-in-beijing-as-death-toll-jumps.
20 Christiaan De Beukelaer, 'Bored of Your Box Room? Try Being Marooned on the Ocean!' *Times Higher Education*, May 4, 2020, www.timeshighereducation.com/opinion/bored-your-box-room-try-being-marooned-ocean.
21 Edward O. Wilson, *Half Earth* (New York: Liveright, 2016).
22 Ole Bjerg, *Parallax of Growth: The Philosophy of Ecology and Economy* (Cambridge; Malden, MA: Polity, 2016).
23 Ghosh, *The Great Derangement*.
24 Ulrich Beck, *World at Risk* (Cambridge; Malden, MA: Polity Press, 2014).
25 This is an excerpt from our collective 'crew statement' made upon arrival in Pointe-à-Pitre. https://christiaan.debeukelaer.net/sailing/safe-aboard-avontuur/.
26 Christiaan De Beukelaer, 'COVID-19 at Sea: "The World as You Know it No Longer Exists,"' *Cultural Studies* 35, no. 2–3 (May 4, 2021): 572–84, https://doi.org/10.1080/09502386.2021.1898020.
27 Ship chandlers operate portside shops that supply ships with food and almost anything else one could need at sea. As ships' time in port is usually limited and expensive, shipowners gladly pay for the convenience of getting goods delivered to the quay. Our manual cargo operations took more time, which would have allowed Cook to go ashore and source cheaper – and better – food herself.
28 Eric Newby, *The Last Grain Race* (London: Secker & Warburg, 1956), 51.

Chapter 4 Coffee, rum, and chocolate

1 The yellow flag serves as a message from ship to shore to request entry into port. As long as the flag is up, the ship and its crew are pending inspections to be 'cleared in.' Once all paperwork and medical checks have been completed, which of course included tests for the SARS-CoV-2 virus in 2020, the flag can come down as a sign that the crew are free from contagious disease. This clearance is hence called 'free pratique.'
2 Marc Levinson, *The Box: How the Shipping Container Made the World Smaller and the World Economy Bigger*, Second edition (Princeton, NJ: Princeton University Press, 2016).
3 Bill Sharpsteen, *The Docks* (Berkeley, CA: University of California Press, 2011).
4 George, *Ninety Percent of Everything*, 17.
5 George, *Ninety Percent of Everything*, 7.

6 Hege Høyer Leivestad and Elisabeth Schober, 'Politics of Scale: Colossal Containerships and the Crisis in Global Shipping,' *Anthropology Today* 37, no. 3 (June 2021): 3–7, https://doi.org/10.1111/14678322.12650.
7 World Health Organization, https://covid19.who.int/region/amro/country/hn.
8 MOL, 'MOL, Tohoku Electric Power Sign Deal for Transport Using Coal Carrier Equipped with Hard Sail Wind Power Propulsion System (Wind Challenger)' (Mitsui O.S.K. Lines, December 10, 2020), www.mol.co.jp/en/pr/2020/img/20085.pdf.
9 Bullock, Mason, and Larkin, 'The Urgent Case for Stronger Climate Targets for International Shipping.'
10 Stopford, *Maritime Economics*.
11 Ministry of Land, Infrastructure, Transport and Tourism, www.mlit.go.jp/en/report/press/kaiji07_hh_000001.html.
12 InfluenceMap, 'Decision Time for the IMO on Climate: The Polarized Struggle among States for Ambitious Climate Policy on Shipping' (London: InfluenceMap, 2018), https://influencemap.org/site/data/000/309/IMO_Shipping_Report_April_2018.pdf.
13 Declan Bush, 'Japan Backs Call to Make Shipping Net Zero by 2050,' *Lloyd's List*, October 26, 2021, https://lloydslist.maritimeintelligence.informa.com/LL1138623/Japan-backs-call-to-make-shipping-net-zero-by-2050.
14 UNCTAD, *Review of Maritime Transport*, 2020, 12.
15 Australian Government, 'The Plan to Deliver Net Zero: The Australian Way' (Canberra: Australian Government, 2021), 17, www.industry.gov.au/sites/default/files/October%202021/document/the-plan-to-deliver-net-zero-the-australian-way.pdf.
16 Australian Government, 'Australia's Long-Term Emissions Reduction Plan: A Whole-of-Economy Plan to Achieve Net Zero Emissions by 2050' (Canberra: Australian Government, 2020), 19, www.industry.gov.au/sites/default/files/October%202021/document/australias-long-term-emissions-reduction-plan.pdf.
17 Stockholm Environment Institute et al., 'The Production Gap: Governments' Planned Fossil Fuel Production Remains Dangerously out of Sync with Paris Agreement Limits' (Stockholm: Stockholm Environment Institute, 2021), http://productiongap.org/2021report.
18 Nick Blenkey, 'Eastern Pacific: "We Won't Carry Coal,"' *MarineLog*, January 25, 2022, www.marinelog.com/legal-safety/environment/eastern-pacific-we-wont-carry-coal/.
19 Donella H. Meadows et al., *The Limits to Growth* (New York: Universe Books, 1972).
20 Graham M. Turner, 'A Comparison of The Limits to Growth with 30 Years of Reality,' *Global Environmental Change* 18, no. 3 (August 2008): 397–411, https://doi.org/10.1016/j.gloenvcha.2008.05.001.

21 Gaya Herrington, 'Update to Limits to Growth: Comparing the World3 Model with Empirical Data,' *Journal of Industrial Ecology* 25, no. 3 (June 2021): 614, https://doi.org/10.1111/jiec.13084.
22 Andrew Curry, 'Revisiting The Limits to Growth,' *The Next Wave* (blog), August 13, 2021, https://thenextwavefutures.wordpress.com/2021/08/13/revisiting-the-limits-to-growth-systems-climate-change/.
23 Harry Dempsey and Dave Lee, 'Amazon, Ikea and Unilever Commit to Zero-Emission Shipping by 2040,' *Financial Times*, October 19, 2021, https://amp.ft.com/content/850eee4b-2c2d-4186-99d7-fdbe8131dddo.
24 Vaclav Smil, *Growth: From Microorganisms to Megacities* (Cambridge, MA: The MIT Press, 2019), 511.
25 Kate Lance, *Alan Villiers: Voyager of the winds* (London: National Maritime Museum, 2009), 8.
26 *The Economist*, 'Accelerating Energy Innovation for the Blue Economy' (*The Economist*, 2020), www.woi.economist.com/wp-content/uploads/2020/11/AcceleratingEnergyInnovationfortheBlueEconomy.pdf.
27 Stopford, *Maritime Economics*, 24.
28 UNCTAD, *Review of Maritime Transport*, 2020, 4.
29 Teikei, www.teikeicoffee.org/en/project/.
30 Market Lane Coffee, https://marketlane.com.au/pages/coffee.
31 Market Lane Coffee, 'Sustainable, Transparent and Ethical Sourcing,' https://marketlane.com.au/pages/sustainable-and-transparent-sourcing.
32 Teikei, www.teikeicoffee.org/en/project/.
33 Freightos Data, https://fbx.freightos.com/.
34 Searates, www.searates.com/.
35 Fairtrade Foundation, '10 Facts About Fairtrade Coffee,' www.fairtrade.org.uk/media-centre/blog/10-facts-about-fairtrade-coffee/.
36 Statista, 'Coffee Consumption Worldwide from 2012/13 to 2020/21,' www.statista.com/statistics/292595/global-coffee-consumption/.
37 *The Herald*, 'Scotland to China and Back Again … Cod's 10,000-Mile Trip to Your Table,' *The Herald*, www.heraldscotland.com/default_content/12765981.scotland-china-back-cods-10-000-mile-trip-table/.
38 Phil Lasker, Jenya Goloubeva, and Bill Birtles, 'Here's How Australia is Planning to Deal with China's Ban on Foreign Waste,' *ABC News*, December 9, 2017, www.abc.net.au/news/201712-10/china-ban-on-foreign-rubbish-leaves-recycling-industry-in-a-mess/9243184.
39 International Transport Forum, 'ITF Transport Outlook 2021.'
40 Stopford, *Maritime Economics*.
41 Hans Rosling, *Factfulness: Ten Reasons We're Wrong about the World – and Why Things Are Better than You Think* (New York: Flatiron Books, 2018).
42 Herrington, 'Update to Limits to Growth.'
43 UNCTAD, *Review of Maritime Transport*, 2021.

44 Bill McKibben, 'The Happiest Number I've Heard in Ages,' Substack newsletter, *The Crucial Years* (blog), January 7, 2022, https://billmckibben.substack.com/p/the-happiest-number-ive-heard-in?s=r.
45 Maria Sharmina et al., 'Global Energy Scenarios and Their Implications for Future Shipped Trade,' *Marine Policy* 84 (October 2017): 12–21, https://doi.org/10.1016/j.marpol.2017.06.025.
46 Xiao-Tong Wang et al., 'Trade-Linked Shipping CO_2 Emissions,' *Nature Climate Change* 11, no. 11 (November 2021): 945–51, https://doi.org/10.1038/s41558-021-01176-6.
47 Sarah McFarland Taylor, *Ecopiety*, Religion and Social Transformation (New York: NYU Press, 2019).
48 H. Damon Matthews et al., 'An Integrated Approach to Quantifying Uncertainties in the Remaining Carbon Budget,' *Communications Earth & Environment* 2, no. 1 (December 2021): 7, https://doi.org/10.1038/s43247-020-00064-9.
49 Bullock, Mason, and Larkin, 'The Urgent Case for Stronger Climate Targets for International Shipping.'
50 Timothy Morton, *Hyperobjects: Philosophy and Ecology after the End of the World* (Minneapolis, MN: University of Minnesota Press, 2013).
51 See the ship profile drawing at the start of the book for detail of the vessel.

Chapter 5 Point of return

1 Déborah Danowski and Eduardo Batalha Viveiros de Castro, *The Ends of the World* (Cambridge: Polity, 2017).
2 Bruno Latour, *Down to Earth: Politics in the New Climatic Regime* (Cambridge; Medford, MA: Polity, 2018).
3 IMO, 'IMO Echoes Shipping Industry Calls for Governments to Keep Shipping and Supply Chains Open and Grant Special Travel Exemptions to Seafarers in COVID-19 Pandemic,' International Maritime Organization, March 31, 2020, www.imo.org/en/MediaCentre/PressBriefings/Pages/09-seafarers-COVID19.aspx.
4 IMO, 'IMO Welcomes UN Resolution on Keyworker Seafarers,' International Maritime Organization, December 1, 2020, www.imo.org/en/MediaCentre/PressBriefings/pages/44-seafarers-UNGA-resolution.aspx.
5 International Transport Workers' Federation, 'ITF Agrees to Crew Contract Extensions,' International Transport Workers' Federation, March 19, 2020, www.itfglobal.org/en/news/itf-agrees-crew-contract-extensions.
6 Wilhelmsen, 'COVID-19 Global Port Restrictions Map,' www.wilhelmsen.com/ships-agency/campaigns/coronavirus/coronavirus-map/.

7 Christiaan De Beukelaer, 'COVID-19 Border Closures Cause Humanitarian Crew Change Crisis at Sea,' *Marine Policy* 132 (October 2021), https://doi.org/10.1016/j.marpol.2021.104661.
8 De Beukelaer, 'The Hundreds of Thousands of Stranded Maritime Workers Are the Invisible Victims of the Pandemic.'
9 IMO, 'Allow Crew Changes to Resolve Humanitarian Crisis, Insists IMO Secretary-General,' International Maritime Organization, September 7, 2020, www.imo.org/en/MediaCentre/PressBriefings/Pages/26-Allow-crew-changes.aspx.
10 UNCTAD, *Review of Maritime Transport*, 2020.
11 UNCTAD, *Review of Maritime Transport*, 2021.
12 Michelle Griffin, 'Months at Sea with No Internet, Sailing Ship Heads Back to a "Different World,"' *The Age*, May 1, 2020, www.theage.com.au/world/central-america/months-at-sea-with-no-internet-sailing-ship-heads-back-to-a-different-world-20200430-p540si.html.
13 Robert Hassan, *Uncontained: Digital Disconnection and the Experience of Time* (Melbourne: Grattan Street Press, 2019).
14 Jason Jiang, 'What Changes Would You like to Make to the Maritime Labour Convention?' Splash247, September 23, 2020, https://splash247.com/what-changes-would-you-like-to-see-the-maritime-labour-convention/.
15 IMO, 'Internet on Ships a Key to Recruiting and Retaining Seafarers, IMO Symposium Told,' September 25, 2015, https://imopublicsite.azurewebsites.net/en/MediaCentre/PressBriefings/Pages/40-WMD-symposium.aspx.
16 Nihar Herwadkar, 'Pros and Cons of Internet Onboard Ships: A Sailor's Perspective,' *Marine Insight* (blog), September 5, 2019, www.marineinsight.com/life-at-sea/seafaring-internet-onboard-ships-sailors-perspective/.
17 Alan Villiers, *Sons of Sindbad* (London: Hodder & Stoughton, 1940), 21.
18 Laleh Khalili, *Sinews of War and Trade: Shipping and Capitalism in the Arabian Peninsula* (London; New York: Verso, 2020).
19 Deborah Cowen, *The Deadly Life of Logistics: Mapping Violence in Global Trade* (Minneapolis, MN: University of Minnesota Press, 2014).
20 Rhys Berry, 'Wind-Propelled Liner to Guarantee Fuel Savings, Price Certainty, Says Neoline CEO,' *Bunkerspot*, June 7, 2021, www.bunkerspot.com/global/53387-global-wind-propelled-liner-to-guarantee-fuel-savings-price-certainty-says-neoline-ceo#.YL3yH7Ql614.linkedin.
21 Allwright, cited in Nicola Cutcher, 'Winds of Trade: Passage to Zero-Emission Shipping,' *American Journal of Economics and Sociology* 79, no. 3 (May 2020): 976, https://doi.org/10.1111/ajes.12331.

22 Nishatabbas Rehmatulla et al., 'Wind Technologies: Opportunities and Barriers to a Low Carbon Shipping Industry,' *Marine Policy* 75 (January 2017): 217–26, https://doi.org/10.1016/j.marpol.2015.12.021.
23 Wind-Assisted Ship Propulsion, 'Enkhuizen Nautical College Sets Course for Wind-Assisted Ship Propulsion and Sustainable Shipping,' *Wind-Assisted Ship Propulsion*, December 1, 2020, https://northsearegion.eu/wasp/news/enkhuizen-nautical-college-sets-course-for-wind-assisted-ship-propulsion-and-sustainable-shipping/.
24 Campling and Colás, *Capitalism and the Sea*, 1.
25 Philip E. Steinberg, *The Social Construction of the Ocean* (Cambridge: Cambridge University Press, 2001).
26 Koji Sekimizu, 'Future-Ready Shipping Conference 2015, Singapore,' International Maritime Organization, September 28, 2015, www.imo.org/en/MediaCentre/SecretaryGeneral/Pages/FRS-keynote.aspx.
27 John Urry, *Societies beyond Oil: Oil Dregs and Social Futures* (London; New York: Zed Books, 2013).
28 Lloyd's Register and UMAS, *Zero-Emission Vessels 2030. How Do We Get There?* (London: Lloyd's Register & University Maritime Advisory Services, 2017).
29 John Asafu-Adjaye et al., 'An Ecomodernist Manifesto' (Breakthrough Institute, 2015), http://rgdoi.net/10.13140/RG.2.1.1974.0646.
30 Sam Cambers, 'More than a Quarter of All Tonnage under Construction Will Use Alternative Fuels,' Splash247, November 30, 2020, https://splash247.com/more-than-a-quarter-of-all-tonnage-under-construction-will-use-alternative-fuels/.
31 Transport & Environment, 'Methane Escaping from "Green" Gas-Powered Ships Fuelling Climate Crisis – Investigation,' Transport & Environment, April 12, 2022, www.transportenvironment.org/discover/methane-escaping-from-green-gas-powered-ships-fuelling-climate-crisis-investigation/.
32 Shippingwatch, 'Nordic Minister Warns the Age of LNG-Fueled Ships is Over,' March 18, 2022, https://shippingwatch.com/carriers/article13839348.ece.
33 Aldo Chircop and Desai Shan, 'Governance of International Shipping in the Era of Decarbonisation: New Challenges for the IMO?' in *Maritime Law in Motion*, ed. Proshanto K. Mukherjee, Maximo Q. Mejia, and Jingjing Xu, vol. 8, WMU Studies in Maritime Affairs (Cham: Springer International Publishing, 2020), 109, https://doi.org/10.1007/978-3-030-31749-2_6.
34 Vaclav Smil, *Energy and Civilization: A History* (Cambridge, MA: The MIT Press, 2017).
35 Thomas Wiedmann et al., 'Scientists' Warning on Affluence,' *Nature Communications* 11, no. 1 (December 2020): 3107, https://doi.org/10.1038/s4

1467-020-16941-y; European Environment Agency, 'Growth without Economic Growth,' Sustainability Transitions (Copenhagen: European Environment Agency, 2021), www.eea.europa.eu/themes/sustainability-transitions/drivers-of-change/growth-without-economic-growth.

36 Kevin Mallinger and Martin Mergili, 'The Global Iron Industry and the Anthropocene,' *The Anthropocene Review*, December 30, 2020, 1–19, https://doi.org/10.1177/2053019620982332.

37 Sharmina et al., 'Global Energy Scenarios and Their Implications for Future Shipped Trade.'

38 UNCTAD, *Review of Maritime Transport*, 2020; UNCTAD, *Review of Maritime Transport*, 2021.

39 Marine Benchmark, 'Maritime CO2 Emissions,' November 2020, Research Brief (Marine Benchmark, 2020), www.marinebenchmark.com/wp-content/uploads/2020/11/Marine-Benchmark-CO2.pdf.

40 Andrew Hook et al., 'A Systematic Review of the Energy and Climate Impacts of Teleworking,' *Environmental Research Letters* 15, no. 9 (August 21, 2020): 093003, https://doi.org/10.1088/1748-9326/ab8a84.

41 Urry, *Societies beyond Oil*, 191.

42 Urry, *Societies beyond Oil*, 214.

43 Jackson, *Prosperity without Growth.*

44 Urry, *Societies beyond Oil*, 210.

45 Jonathan Symons, *Ecomodernism: Technology, Politics and the Climate Crisis* (Cambridge: Polity Press, 2019).

46 Owen Gaffney and Johan Rockström, *Breaking Boundaries: The Science of Our Planet* (London: DK, 2021); Raworth, *Doughnut Economics*; Jackson, *Prosperity without Growth*; Rockström et al., 'A Safe Operating Space for Humanity'; Parrique, 'The Political Economy of Degrowth.'

47 Klein, *On Fire: The Burning Case for a Green New Deal*; Klein, *This Changes Everything*; Bill McKibben, *Eaarth: Making a Life on a Tough New Planet* (New York: Time Books, 2010).

48 Dana, *Two Years Before the Mast*, 32.

49 Bruno Latour, 'Imaginer les gestes-barrières contre le retour à la production d'avant-crise,' *AOC media – Analyse Opinion Critique* (blog), March 29, 2020, https://aoc.media/opinion/2020/03/29/imaginer-les-gestes-barrieres-contre-le-retour-a-la-production-davant-crise/.

50 Chircop and Shan, 'Governance of International Shipping in the Era of Decarbonisation,' 107.

51 Tom Lemaire, *De Val van Prometheus: Over de Keerzijden van de Vooruitgang* (Amsterdam: Ambo, 2010).

52 Nicola Davison, 'The Anthropocene Epoch: Have We Entered a New Phase of Planetary History?' *The Guardian*, May 30, 2019, sec. Environment, www.theguardian.com/environment/2019/may/30/anthropocene-epoch-have-we-entered-a-new-phase-of-planetary-history.

53 Anthropocene Working Group, 'Results of Binding Vote by AWG,' May 21, 2019, http://quaternary.stratigraphy.org/working-groups/anthropocene/.
54 Asafu-Adjaye et al., 'An Ecomodernist Manifesto'; Erle C. Ellis, *Anthropocene: A Very Short Introduction*, Very Short Introductions (Oxford; New York: Oxford University Press, 2018); Symons, *Ecomodernism.*
55 Corinna Burkhart, Matthias Schmelzer, and Nina Treu, eds, *Degrowth in Movement(s): Exploring Pathways for Transformation* (Winchester; Washington, DC: Zero Books, 2020); Giacomo D'Alisa, Federico Demaria, and Giorgos Kallis, eds, *Degrowth: A Vocabulary for a New Era* (New York: Routledge, 2015); Jason Hickel, *Less is More: How Degrowth Will Save the World* (London: Windmill Books, 2021); Giorgos Kallis, *The Case for Degrowth*, The Case for Series (Cambridge: Polity Press, 2020); Serge Latouche, *Farewell to Growth* (Cambridge; Malden, MA: Polity, 2009); Parrique, 'The Political Economy of Degrowth'; Jackson, *Prosperity without Growth.*

Chapter 6 The eternal frontier

1 Offshore Energy, 'ExxonMobil in Another Extension for Rowan Drillship,' www.offshore-energy.biz/exxonmobil-in-another-extension-for-rowan-drillship/.
2 Bureau of Safety and Environmental Enforcement, 'How Many Platforms are in the Gulf of Mexico?' www.bsee.gov/faqs/how-many-platforms-are-in-the-gulf-of-mexico.
3 National Oceanic and Atmospheric Administration, Gulf of Mexico Data Atlas, www.ncei.noaa.gov/maps/gulf-data-atlas/atlas.htm?plate=Offshore%20Structures.
4 United States Coast Guard, 'On Scene Coordinator Report on Deepwater Horizon Oil Spill' (Washington DC: United States Coast Guard, 2011), https://homeport.uscg.mil/Lists/Content/Attachments/119/DeepwaterHorizonReport%2031Aug2011%20-CD_2.pdf.
5 Jonny Beyer et al., 'Environmental Effects of the Deepwater Horizon Oil Spill: A Review,' *Marine Pollution Bulletin* 110, no. 1 (September 2016): 28–51, https://doi.org/10.1016/j.marpolbul.2016.06.027.
6 MTI Network, 'Torrey Canyon Changes International Maritime Regulations,' nd, www.mtinetwork.com/torrey-canyon-changes-international-maritime-regulations/. Accessed March 15, 2021.
7 Climate Action Tracker, 'Climate Summit Momentum: Paris Commitments Improved Warming Estimate to 2.4°C.'
8 IPCC, 'Climate Change 2021: The Physical Science Basis Summary for Policymakers,' in *Climate Change 2021: The Physical Science Basis. Contribution of Working Group I to the Sixth Assessment Report of the Intergovernmental Panel on Climate Change* (Cambridge: Cambridge University Press, 2021), 38.

9 Fiona Harvey, 'No New Oil, Gas or Coal Development if World is to Reach Net Zero by 2050, Says World Energy Body,' *The Guardian*, May 18, 2021, sec. Environment, www.theguardian.com/environment/2021/may/18/no-new-investment-in-fossil-fuels-demands-top-energy-economist.
10 Graham Readfearn, '"Nothing off Limits": Offshore Gas and Oil Exploration Area 5km from Twelve Apostles,' *The Guardian*, June 15, 2021, sec. Australia news, www.theguardian.com/australia-news/2021/jun/15/nothing-off-limits-offshore-gas-and-oil-exploration-area-5km-from-twelve-apostles.
11 Australian Government, 'The Plan to Deliver Net Zero: The Australian Way.'
12 Adam Morton and Bec Pridham, 'Australia Considering More than 100 Fossil Fuel Projects That Could Produce 5% of Global Industrial Emissions,' *The Guardian*, November 2, 2021, sec. Environment, www.theguardian.com/environment/2021/nov/03/australia-considering-more-than-100-fossil-fuel-projects-that-could-produce-5-of-global-industrial-emissions.
13 Department of Industry, Science, Energy and Resources, 'National Greenhouse Gas Inventory Quarterly Update: June 2021' (Canberra: Australian Government, 2021), www.industry.gov.au/data-and-publications/national-greenhouse-gas-inventory-quarterly-update-june-2021.
14 Sekula and Burch, *The Forgotten Space*.
15 Hage, *Is Racism an Environmental Threat?*, 49.
16 Helen S. Findlay and Carol Turley, 'Ocean Acidification and Climate Change,' in *Climate Change*, ed. Trevor M. Letcher (Elsevier, 2021), 251, https://doi.org/10.1016/B978-0-12-821575-3.00013-X.
17 Lijing Cheng et al., 'Upper Ocean Temperatures Hit Record High in 2020,' *Advances in Atmospheric Sciences* 38, no. 4 (April 2021): 523–30, https://doi.org/10.1007/s00376-021-0447-x.
18 Niklas Boers, 'Observation-Based Early-Warning Signals for a Collapse of the Atlantic Meridional Overturning Circulation,' *Nature Climate Change* 11, no. 8 (August 2021): 680–8, https://doi.org/10.1038/s41558-021-01097-4.
19 James Lovelock, 'Beware: Gaia May Destroy Humans before We Destroy the Earth,' *The Guardian*, November 2, 2021, sec. Opinion, www.theguardian.com/commentisfree/2021/nov/02/beware-gaia-theory-climate-crisis-earth.
20 Campling and Colás, *Capitalism and the Sea*, 16.
21 Klein, *This Changes Everything*, 20–1.
22 Ghosh, *The Nutmeg's Curse*.
23 Ghosh, *The Nutmeg's Curse*, 167.
24 IPCC, 'Climate Change 2022: Impacts, Adaptation and Vulnerability' (Intergovernmental Panel on Climate Change, 2022), various pages.

25 Ėtkind, *Nature's Evil.*
26 Ėtkind, *Nature's Evil.*
27 IMO, 'Member States, IGOs and NGOs,' www.imo.org/en/About/Membership/Pages/Default.aspx.
28 IMO, 'IMO Assembly Elects 40-Member Council,' www.imo.org/en/MediaCentre/PressBriefings/Pages/32-Council-elections-A31.aspx.
29 Jack Corbett et al., 'Climate Governance, Policy Entrepreneurs and Small States: Explaining Policy Change at the International Maritime Organisation,' *Environmental Politics* 29, no. 5 (July 28, 2020): 825–44, https://doi.org/10.1080/09644016.2019.1705057.
30 See DeSombre, *Flagging Standards.*
31 Matt Apuzzo and Sarah Hurtes, 'Tasked to Fight Climate Change, a Secretive U.N. Agency Does the Opposite,' *New York Times*, June 3, 2021, www.nytimes.com/2021/06/03/world/europe/climate-change-un-international-maritime-organization.html; Transparency International, *Governance at the International Maritime Organization.*
32 Eoin Bannon, 'UN Shipping Agency Greenlights a Decade of Rising Greenhouse Gas Emissions,' *Transport & Environment*, November 17, 2020, www.transportenvironment.org/press/un-shipping-agency-greenlights-decade-rising-greenhouse-gas-emissions; Ocean Conservancy, 'UN Shipping Agency Greenlights a Decade of Rising Greenhouse Gas Emissions,' *Ocean Conservancy*, November 17, 2020, https://oceanconservancy.org/news/un-shipping-agency-greenlights-decade-rising-greenhouse-gas-emissions/; Rutherford, Mao, and Comer, 'Potential CO_2 Reductions under the Energy Efficiency Existing Ship Index.'
33 Bullock et al., 'Shipping and the Paris Climate Agreement,' 14.
34 Bullock, Mason, and Larkin, 'The Urgent Case for Stronger Climate Targets for International Shipping.'
35 Transparency International, *Governance at the International Maritime Organization.*
36 Lucy Gilliam, 'World Leaders Demand Shipping Emissions Cuts under Countries' Paris Climate Commitments,' Transport & Environment, December 13, 2017, www.transportenvironment.org/discover/world-leaders-demand-shipping-emissions-cuts-under-countries-paris-climate-commitments/.
37 Chloé Farand, 'Anger as UN Body Approves Deal That Allows Ship Emissions to Rise to 2030,' Climate Home News, November 17, 2020, www.climatechangenews.com/2020/11/17/anger-un-body-approves-deal-allows-ship-emissions-rise-2030/.
38 Adamopoulos, 'Marshall Islands Demands Carbon Pricing Measure from the IMO.'

39 Thomas G. Weiss, *Would the World Be Better without the UN?* (Cambridge: Polity Press, 2018).
40 Heleen L. van Soest, Michel G. J. den Elzen, and Detlef P. van Vuuren, 'Net-Zero Emission Targets for Major Emitting Countries Consistent with the Paris Agreement,' *Nature Communications* 12, no. 1 (December 2021): 2140, https://doi.org/10.1038/s41467-021-22294-x; IPCC, 'The Evidence is Clear: The Time for Action is Now. We Can Halve Emissions by 2030. | UNFCCC,' 2022, https://unfccc.int/news/the-evidence-is-clear-the-time-for-action-is-now-we-can-halve-emissions-by-2030.
41 Ghosh, *The Great Derangement*, 160.
42 Nathaniel Rich, *Losing Earth: The Decade We Could Have Stopped Climate Change* (London: Picador, 2020).
43 Climate action needs to respect two simple rules of politics. First, there is no victory in avoiding something; victory lies in doing something. It's far easier to win an election based on building something, anything, new than it is for avoiding risk. The reason is that claiming credit for the former is easier than taking responsibility for the latter. Second, regulation must always be economically viable. Gavin Allwright told me that the IMO adopted a clear carbon emissions target once the industry was convinced it would be economically feasible. As David Whyte argues in his book *Ecocide*, environmental regulation does not aim to stop pollution; it merely sets upper limits for corporate pollution. What those limits are and how to balance them with a concern for future life on the planet is what is at stake.
44 Peter Christoff, 'Net Zero by 2050? Even if Scott Morrison Gets the Nationals on Board, Hold the Applause,' *The Conversation*, June 23, 2021, http://theconversation.com/net-zero-by-2050-even-if-scott-morrison-gets-the-nationals-on-board-hold-the-applause-163074.
45 James Dyke, Robert Watson, and Wolfgang Knorr, 'Climate Scientists: Concept of Net Zero is a Dangerous Trap,' *The Conversation*, April 22, 2021, http://theconversation.com/climate-scientists-concept-of-net-zero-is-a-dangerous-trap-157368.
46 Rehmatulla et al., 'Wind Technologies.'
47 www.youtube.com/watch?v=ZNaAfgZrLZU.
48 Rich, *Losing Earth*, 98.
49 Anna Lowenhaupt Tsing, *Friction: An Ethnography of Global Connection* (Princeton, NJ: Princeton University Press, 2005), 84.
50 Timbercoast, https://timbercoast.com/en/.
51 Fairtransport's mission statement, https://fairtransport.eu/about/.
52 www.sailcargo.inc/en/home-1.
53 IMO, 'Fourth IMO Greenhouse Gas Study,' Annex 1, page 3.
54 Andrew Simons, 'Cargo Sailing: A Life Cycle Assessment Case Study' (Grasswil & Den Helder: 3SP Switzerland & EcoClipper, 2020).

55 Kris De Decker, 'How to Design a Sailing Ship for the 21st Century?' *Low←Tech Magazine*, May 11, 2021, https://solar.lowtechmagazine.com/2021/05/how-to-design-a-sailing-ship-for-the-21st-century.html.
56 Donna Jeanne Haraway, 'Anthropocene, Capitalocene, Plantationocene, Chthulucene: Making Kin,' *Environmental Humanities* 6, no. 1 (2015): 159, https://doi.org/10.1215/22011919-3615934.
57 Susan Smillie, 'Long Journey Home: The Stranded Sailboats in a Race to Beat the Hurricanes,' *The Guardian*, May 12, 2020, www.theguardian.com/environment/2020/may/12/long-journey-home-the-stranded-sailboats-in-a-race-to-beat-the-hurricanes.
58 CBC, 'Overboard,' www.cbc.ca/player/play/1367419135.
59 Dana, *Two Years Before the Mast*, 376.

Chapter 7 Ship Earth

1 www.youtube.com/watch?v=ooIxHVXgLbc.
2 Robert Macfarlane, *Underland: A Deep Time Journey*, First American edition (New York: W. W. Norton & Company, 2019).
3 Ghosh, *The Great Derangement*, 80.
4 Naomi Oreskes and Erik M. Conway, *The Collapse of Western Civilisation: A View from the Future* (New York: Columbia University Press, 2014), 38.
5 Tim Jackson, *Post Growth: Life After Capitalism* (Cambridge: Polity, 2021), 3.
6 IMO, 'Fourth IMO Greenhouse Gas Study'; IMO, 'Third IMO Greenhouse Gas Study.'
7 Sekimizu, 'Future-Ready Shipping Conference 2015, Singapore.'
8 IMO, 'Third IMO Greenhouse Gas Study.'
9 Naomi Oreskes and Erik M. Conway, *Merchants of Doubt: How a Handful of Scientists Obscured the Truth on Issues from Tobacco Smoke to Global Warming* (New York: Bloomsbury Press, 2011).
10 UNCTAD, *Review of Maritime Transport*, 2020, 44.
11 George, *Ninety Percent of Everything*, 245.
12 DeSombre, *Flagging Standards*.
13 Bangstad and Nilsen, 'Thoughts on the Planetary.'
14 Tim Gore, 'Confronting Carbon Inequality: Putting Climate Justice at the Heart of the COVID-19 Recovery' (Nairobi: Oxfam, 2020), www.oxfam.org/en/research/confronting-carbon-inequality.
15 Dario Kenner, *Carbon Inequality: The Role of the Richest in Climate Change* (Abingdon: Routledge, 2019).
16 Marshall Islands and Solomon Islands, 'Proposal for IMO to Establish a Universal Mandatory Greenhouse Gas Levy.'

17 Hilda Heine and Christiana Figueres, 'Polluters on the High Seas,' *New York Times*, April 6, 2018, www.nytimes.com/2018/04/06/opinion/greenhouse-gases-international-shipping.html.
18 Heine and Figueres, 'Polluters on the High Seas'.
19 Farhana Yamin, 'The High Ambition Coalition,' in *Negotiating the Paris Agreement: The Insider Stories*, ed. Henrik Jepsen et al. (Cambridge; New York: Cambridge University Press, 2021), 237.
20 Yamin, 'The High Ambition Coalition,' 238–9.
21 Climate Action Tracker, 'Climate Summit Momentum: Paris Commitments Improved Warming Estimate to 2.4°C.'
22 Malte Meinshausen et al., 'Realization of Paris Agreement Pledges May Limit Warming Just below 2 °C,' *Nature* 604, no. 7905 (April 14, 2022): 308–9, https://doi.org/10.1038/s41586-022-04553-z.
23 Christopher Lyon et al., 'Climate Change Research and Action Must Look beyond 2100,' *Global Change Biology*, September 24, 2021, 1, https://doi.org/10.1111/gcb.15871.
24 Joshua Mcdonald, 'Rising Sea Levels Threaten Marshall Islands' Status as a Nation, World Bank Report Warns,' *The Guardian*, October 16, 2021, sec. World news, www.theguardian.com/world/2021/oct/17/rising-sea-levels-threaten-marshall-islands-status-as-a-nation-world-bank-report-warns.
25 David Freestone and Duygu Çiçek, 'Legal Dimensions of Sea Level Rise: Pacific Perspectives' (Washington DC: World Bank, 2021), https://openknowledge.worldbank.org/handle/10986/35881.
26 Maryam Omidi, 'Maldives Sends Climate SOS with Undersea Cabinet,' *Reuters*, October 17, 2009, www.reuters.com/article/us-maldives-environment-idUSTRE59G0P120091017.
27 Tony de Brum, 'Minister de Brum of Marshall Islands Calls for Massive Global Response to Climate Crisis,' *Climate & Development Knowledge Network*, 2015, https://cdkn.org/2015/10/news-minister-de-brum-of-marshall-islands-calls-for-global-crisis-response-to-climate-change/?loclang=en_gb.
28 Joseph Earsom and Tom Delreux, 'A Nice Tailwind: The EU's Goal Achievement at the IMO Initial Strategy,' *Politics and Governance* 9, no. 3 (September 30, 2021): 401–11, https://doi.org/10.17645/pag.v9i3.4296.
29 Corbett et al., 'Climate Governance, Policy Entrepreneurs and Small States.'
30 www.register-iri.com/maritime/international-participation/imo/.
31 Corbett et al., 'Climate Governance, Policy Entrepreneurs and Small States,' 833.
32 One Planet Summit. 2017. 'Tony de Brum Declaration.'

33 Fiona Harvey, 'Tony de Brum Obituary,' *The Guardian*, October 10, 2017, sec. Environment, www.theguardian.com/environment/2017/oct/10/tony-de-brum-obituary.

34 Corbett et al., 'Climate Governance, Policy Entrepreneurs and Small States,' 837.

35 Fiona Harvey, 'UN Chief Urges Airlines and Shipping Firms to Do More to Cut Emissions,' *The Guardian*, October 14, 2021, sec. Environment, www.theguardian.com/environment/2021/oct/14/un-chief-urges-airlines-and-shipping-firms-to-do-more-to-cut-emissions.

36 Marshall Islands and Solomon Islands, 'Proposal for IMO to Establish a Universal Mandatory Greenhouse Gas Levy.'

37 Carsten Ned Nemra, 'Global Shipping is a Big Emitter, the Industry Must Commit to Drastic Action before it is Too Late,' *The Guardian*, September 19, 2021, sec. World news, www.theguardian.com/world/2021/sep/20/global-shipping-is-a-big-emitter-the-industry-must-commit-to-drastic-action-before-it-is-too-late.

38 World Bank, 'CO2 Emissions (Metric Tons Per Capita),' https://data.worldbank.org/indicator/EN.ATM.CO2E.PC?most_recent_value_desc=true.

39 Akenji et al., *1.5-Degree Lifestyles: Towards A Fair Consumption Space for All*, 42.

40 We made 165 nautical miles on June 18–19 (noon to noon), during a 23h day. The day after, we managed 175 nautical miles, though, as June 19–20 was a 24h day, during which we did not have to adjust our clocks because of the eastward passage; we made some 7.1 and 7.2 knots respectively.

41 Danowski and Castro, *The Ends of the World*; De Beukelaer, 'Tack to the Future.'

42 IPCC, 'Climate Change 2022: Mitigation of Climate Change (Summary for Policymakers),' 41.

43 Timothée Parrique, 'Sufficiency Means Degrowth,' 2022, https://timotheeparrique.com/sufficiency-means-degrowth/.

44 Colin Schultz, 'Shackleton Probably Never Took Out an Ad Seeking Men for a Hazardous Journey,' *Smithsonian Magazine*, www.smithsonianmag.com/smart-news/shackleton-probably-never-took-out-an-ad-seeking-men-for-a-hazardous-journey-5552379/.

45 See, for example, Joe Russell, *The Last Schoonerman: The Remarkable Life of Captain Lou Kenedy* (Rockledge FL: Nautical Publishing Company, 2006).

46 Ken McCulloch, 'Sail Training,' in *Routledge International Handbook of Outdoor Studies*, ed. Barbara Humberstone, Heather Prince, and Karla A. Henderson (New York: Routledge, 2015), 236–43, https://doi.org/10.4324/9781315768465-27; Manu Schijf, Pete Allison, and Kris Von Wald, 'Sail Training: A Systematic Review,' *Journal of Outdoor Recreation,*

Education, and Leadership 9, no. 2 (2017): 167–80, https://doi.org/10.18666/JOREL-2017-V9-I2-8230.

47 Bark Europa, 'What We Do,' www.barkeuropa.com/what-we-do.

48 Jean-Paul Rodrigue, 'Fuel Consumption by Containership Size and Speed,' The Geography of Transport Systems, 2020, https://transportgeography.org/contents/chapter4/transportation-and-energy/fuel-consumption-containerships/.

49 Mack, *The Sea*, 160.

50 Campling and Colás, *Capitalism and the Sea*, 14–15.

51 Lovelock, *Gaia*.

52 Alan Weisman, *The World Without Us* (London: Virgin Books, 2004).

53 Davor Vidas, 'The Law of the Sea for a New Epoch?' in *Tidalectics: Imagining an Oceanic Worldview through Art and Science*, ed. Stefanie Hessler (London; Cambridge, MA: TBA21-Academy; The MIT Press, 2018), 236.

54 Richard Sennett, *Together: The Rituals, Pleasures and Politics of Cooperation* (London: Penguin Books, 2013), 9.

55 Sennett, *Together*, 30.

Chapter 8 Where are we headed?

1 Cowen, *The Deadly Life of Logistics*.

2 C. De Beukelaer and J. Corcoran, 'In the Stillness Between Two Waves of the Sea,' *Ocean Archive*, May 7, 2020, https://ocean-archive.org/collection/45.

3 Climate Action Tracker, 'Climate Action Tracker,' 2022, https://climateactiontracker.org/.

4 Steffen et al., 'Trajectories of the Earth System in the Anthropocene,' 8257.

5 Rich, *Losing Earth*.

6 Klein, *This Changes Everything*.

7 Yuval Noah Harari, *Sapiens: A Brief History of Humankind* (New York: Random House, 2015).

8 Geoffroy Delorme, *L'homme-chevreuil: sept ans de vie sauvage* (Paris: les Arènes, 2021).

9 www.youtube.com/watch?v=07EpiXViSIQ.

10 www.youtube.com/watch?v=07EpiXViSIQ.

11 Union of Concerned Scientists, 'Climate Science,' www.ucsusa.org/climate/science.

12 BBC, 'John Kerry Says Glasgow COP26 is the Last Best Hope for the World,' *BBC News*, October 19, 2021, sec. Scotland, www.bbc.com/news/uk-scotland-58914524.

13 IPCC, 'The Evidence is Clear: The Time for Action is Now.'
14 World Commission on Environment and Development, *Our Common Future*, Oxford Paperbacks (Oxford; New York: Oxford University Press, 1987).
15 Kate Crowley, 'Is Australia Faking It? The Kyoto Protocol and the Greenhouse Policy Challenge,' *Global Environmental Politics* 7, no. 4 (November 2007): 118–39, https://doi.org/10.1162/glep.2007.7.4.118; Clive Hamilton and Lins Vellen, 'Land-Use Change in Australia and the Kyoto Protocol,' *Environmental Science & Policy* 2, no. 2 (May 1999): 145–52, https://doi.org/10.1016/S1462-9011(99)00007-6.
16 UNFCCC, 'Glasgow Climate Pact,' Conference of the Parties Serving as the Meeting of the Parties to the Paris Agreement (Glasgow: UNFCCC, 2021), 2, https://unfccc.int/documents/311127.
17 UNFCCC, 'Glasgow Climate Pact,' 4–5.
18 Melissa Clarke, 'COP26 Agreement to Phase Down Coal Not "a Death Knell" for Coal Power says PM, Disputing Boris Johnson,' *ABC News*, November 15, 2021, www.abc.net.au/news/2021-11-15/cop26-phase-down-coal-agreement-not-end-of-industry-says-pm/100621242.
19 Damian Carrington, '"Blah, Blah, Blah": Greta Thunberg Lambasts Leaders over Climate Crisis,' *The Guardian*, September 28, 2021, sec. Environment, www.theguardian.com/environment/2021/sep/28/blah-greta-thunberg-leaders-climate-crisis-co2-emissions.
20 Jonathan Franzen, *What If We Stopped Pretending?* (London: 4th Estate, 2021).
21 Andreas Malm, *The Progress of This Storm: Nature and Society in a Warming World* (London: Verso, 2017), Chapter 8.
22 Malm, *Fossil Capital*, Chapter 8.
23 Apuzzo and Hurtes, 'Tasked to Fight Climate Change, a Secretive U.N. Agency Does the Opposite.'
24 Declan Bush, 'EU States Block Zero-Emissions Resolution at IMO,' *Lloyd's List*, November 22, 2021, https://lloydslist.maritimeintelligence.informa.com/LL1138950/EU-states-block-zero-emissions-resolution-at-IMO.
25 Declan Bush, 'Little Time to Talk Carbon Taxes as IMO States Wrestle with 2050 Resolution,' *Lloyd's List*, November 23, 2021, https://lloydslist.maritimeintelligence.informa.com/LL1138970/Little-time-to-ta lk-carbon-taxes-as-IMO-states-wrestle-with-2050-resolution.
26 James Lovelock, *The Revenge of Gaia* (London: Allan Lane, 2006), 152.
27 Christiaan De Beukelaer, 'Feeling Flight Shame? Try Quitting Air Travel and Catch a Sail Boat,' *The Conversation*, October 2, 2019, https://theconversation.com/feeling-flight-shame-try-quitting-air-travel-and-catch-a-sail-boat-123349.

28 Both virtue ethics and consequentialism are valid rationales for ethical action, though it's unfair to judge actions inspired by one principle if that very action is inspired by another principle.
29 Global Monitoring Laboratory, 'Mauna Loa CO2 Monthly Mean Data.'
30 Rogelj et al., 'Mitigation Pathways Compatible with 1.5°C in the Context of Sustainable Development.'
31 International Energy Agency, 'Global CO2 Emissions Rebounded to Their Highest Level in History in 2021,' IEA, March 8, 2022, www.iea.org/news/global-co2-emissions-rebounded-to-their-highest-level-in-history-in-2021.
32 Alan Evans et al., 'Cop26 President Declares "Fragile Win" for Climate despite Watered-down Coal Pledges – as It Happened,' *The Guardian*, November 14, 2021, sec. Environment, www.theguardian.com/environment/live/2021/nov/13/cop26-live-third-draft-text-expected-as-climate-talks-go-into-overtime.
33 ICS, 'International Chamber of Shipping Sets out Plans for Global Carbon Levy to Expedite Industry Decarbonisation,' International Chamber of Shipping, September 6, 2021, www.ics-shipping.org/press-release/international-chamber-of-shipping-sets-out-plans-for-global-carbon-levy/.
34 Coalition for Zero Emission Vessels, 'Leading Cargo Owners Stand Together for Maritime Decarbonization: Our Shared Ambition and Call for Policy Action' (Aspen Institute, 2021), www.cozev.org/img/FINAL-coZEV-2040-Ambition-Statement_2021-10-18-144834_uorz.pdf.
35 Transport & Environment, 'Methane Escaping from "Green" Gas-Powered Ships Fuelling Climate Crisis – Investigation.'
36 Ship it Zero, 'Big Retail Fossil-Free Shipping Commitment Historic, But Too Weak,' ShipitZero, October 19, 2021, https://shipitzero.org/climate-advocates-big-retail-fossil-free-shipping-commitment-historic-but-too-weak2/.
37 Mærsk Mc-Kinney Møller Center for Zero Carbon Shipping, 'We Show the World it is Possible: Industry Transition Strategy October 2021' (Copenhagen: Mærsk Mc-Kinney Møller Center for Zero Carbon Shipping, October 2021), 30.
38 The World Bank, 'Carbon Revenues from International Shipping: Enabling an Effective and Equitable Energy Transition' (Washington DC: International Bank for Reconstruction and Development / The World Bank, 2022).
39 Bullock, Mason, and Larkin, 'The Urgent Case for Stronger Climate Targets for International Shipping.'
40 Peder Osterkamp, Tristan Smith, and Kasper Søgaard, 'Five Percent Zero Emission Fuels by 2030 Needed for Paris-Aligned Shipping Decarbonization' (Getting to Zero Coalition, 2021),

www.globalmaritimeforum.org/content/2021/03/Getting-to-Zero-Coalition_Five-percent-zero-emission-fuels-by-2030.pdf.

41 Rehmatulla et al., 'Wind Technologies,' 218.

42 Berrill, 'Sail Cargo Reveals Plans for Six 100-TEU Emission-Free Ships.'

43 The Maritime Executive, 'Unique Sail Cargo Ships Departs on First Atlantic Crossing From France,' The Maritime Executive, November 20, 2020, www.maritime-executive.com/article/unique-sail-cargo-ships-departs-on-first-atlantic-crossing-from-france.

44 *Le Figaro*, 'La Bretagne lance une filière de transport maritime à propulsion par le vent,' *Le Figaro*, November 10, 2021, www.lefigaro.fr/flash-eco/la-bretagne-lance-une-filiere-de-transport-maritime-a-propulsion-par-le-vent-20211110.

45 International Transport Workers' Federation, 'UN Global Compact and Shipping Industry Confirm Formation of "People-Centred" Task Force to Ensure Just Transition to Net-Zero,' International Transport Workers' Federation, November 10, 2021, www.itfglobal.org/en/news/un-global-compact-and-shipping-industry-confirm-formation-people-centred-task-force-ensure.

46 Bullock, Mason, and Larkin, 'The Urgent Case for Stronger Climate Targets for International Shipping.'

47 Jason Monios and Adolf K. Y. Ng, 'Competing Institutional Logics and Institutional Erosion in Environmental Governance of Maritime Transport,' *Journal of Transport Geography* 94 (June 2021): 103114, https://doi.org/10.1016/j.jtrangeo.2021.103114.

48 Transport & Environment, 'Planes and Ships Can't Escape Paris Climate Commitments,' May 4, 2018, www.transportenvironment.org/discover/planes-and-ships-cant-escape-paris-climate-commitments/.

49 Transport & Environment, 'Don't Sink Paris: Legal Basis for Inclusion of Aviation and Shipping Emissions in Paris Targets' (Brussels: Transport & Aviation, 2021), www.transportenvironment.org/wp-content/uploads/2021/10/Briefing-paper-NDCs-legal-advice-Aviation-Shipping-Final-2021-2.pdf.

50 Laura Cole, 'COP26: Commission Official Says Maritime Emissions Belong in NDCs,' ENDS Europe, November 2021, www.endseurope.com/article/1733097?utm_source=website&utm_medium=social.

51 Henrik Selin et al., 'Mitigation of CO_2 Emissions from International Shipping through National Allocation,' *Environmental Research Letters* 16, no. 4 (April 1, 2021): 045009, https://doi.org/10.1088/1748-9326/abec02.

52 Garrett Hardin, 'The Tragedy of the Commons,' *Science* 162, no. 3859 (1968): 1243–8.

53 Elinor Ostrom, *Governing the Commons: The Evolution of Institutions for Collective Action* (Cambridge: Cambridge University Press, 1990).

54 Malm, *The Progress of This Storm: Nature and Society in a Warming World*, Chapter 8.
55 Akenji et al., *1.5-Degree Lifestyles: Towards A Fair Consumption Space for All*, 13.
56 Akenji et al., *1.5-Degree Lifestyles: Towards A Fair Consumption Space for All*, 18.
57 Akenji et al., *1.5-Degree Lifestyles: Towards A Fair Consumption Space for All*, 42.
58 David Wallace-Wells, *The Uninhabitable Earth: A Story of the Future* (London: Penguin Books, 2019), 1.
59 Klein, *This Changes Everything*, 62. Emphasis in original.
60 Ghosh, *The Nutmeg's Curse*, 170.
61 Malm, *The Progress of This Storm: Nature and Society in a Warming World*, Chapter 8.
62 Elizabeth R. DeSombre, *Why Good People Do Bad Environmental Things* (Oxford: Oxford University Press, 2018).
63 Lauren Etter and Brendan Murray, 'Shipping Companies Had a $150 Billion Year. Economists Warn They're Also Stoking Inflation,' Bloomberg, January 18, 2022, www.bloomberg.com/news/features/2022-01-18/supply-chain-crisis-helped-shipping-companies-reap-150-billion-in-2021.
64 Bruno Latour, *After Lockdown*, trans. Julie Rose (Cambridge: Polity Press, 2021), 54.

Chapter 9 Sailing home

1 Victor W. Turner, *The Ritual Process: Structure and Anti-Structure*, The Lewis Henry Morgan Lectures 1966 (New York: Aldine de Gruyter, 1995).
2 Mike Hulme, 'One Earth, Many Futures, No Destination,' in *Negotiating Climate Change in Crisis*, ed. Steffen Böhm and Sian Sullivan (Cambridge: Open Book Publishers, 2021), 3–11.
3 Tsing, *Friction*, 85.
4 Latour, *Down to Earth*, 2. Emphasis in original.
5 Ulrich Beck, *The Metamorphosis of the World* (Cambridge; Malden, MA: Polity, 2016).
6 Hulme, 'One Earth, Many Futures, No Destination.'
7 Asafu-Adjaye et al., 'An Ecomodernist Manifesto.'
8 Hulme, 'One Earth, Many Futures, No Destination,' 6; Klein, *This Changes Everything*.
9 Ann Pettifor, *The Case for the Green New Deal* (London: Verso, 2019).
10 Danowski and Castro, *The Ends of the World*.
11 Ghosh, *The Great Derangement*, 128–9. Emphasis added.
12 Simon L. Lewis and Mark A. Maslin, *The Human Planet: How We Created the Anthropocene* (London: Pelican Books, 2018).

13 Lewis and Maslin, *The Human Planet: How We Created the Anthropocene*, 318.
14 Lewis and Maslin, *The Human Planet: How We Created the Anthropocene*, 348–9.
15 Gaffney and Rockström, *Breaking Boundaries*.
16 UNFCCC, 'Glasgow Climate Pact,' 4–5.
17 Herman Melville, *Moby-Dick; Or, The Whale* (New York: Harper & Brothers, 1851), Chapter 9.
18 UNCTAD, *Review of Maritime Transport*, 2021, 13.
19 See Renisa Mawani, 'The Ship, the Slave, the Legal Person,' in *Studies in Law, Politics, and Society*, ed. Austin Sarat, George Pavlich, and Richard Mailey (Bingley: Emerald Publishing Limited, 2022), 19–42, https://doi.org/10.1108/S1059-433720220000087B002 for a detailed discussion of the status of the ship as a 'legal person' at the time when slavery was banned.
20 ICS, 'International Chamber of Shipping Sets out Plans for Global Carbon Levy to Expedite Industry Decarbonisation.'
21 Mærsk Mc-Kinney Møller Center for Zero Carbon Shipping, 'We Show the World it is Possible: Industry Transition Strategy October 2021'; UMAS and Getting to Zero Coalition, 'Closing the Gap: An Overview of the Policy Options to Close the Competitiveness Gap and Enable an Equitable Zero-Emission Fuel Transition in Shipping' (Copenhagen: Getting to Zero Coalition, 2021).
22 Jocelyn Timperley, 'The Broken $100-Billion Promise of Climate Finance – and How to Fix It,' *Nature* 598, no. 7881 (October 21, 2021): 400–2, https://doi.org/10.1038/d41586-021-02846-3.
23 IPCC, 'Climate Change 2022: Mitigation of Climate Change (Full Report)' (Intergovernmental Panel on Climate Change, 2022), Chapter 5, 37–8.
24 Parrique, 'Sufficiency Means Degrowth.'
25 Jackson, *Post Growth: Life After Capitalism*.
26 Paul Balcombe et al., 'How to Decarbonise International Shipping: Options for Fuels, Technologies and Policies,' *Energy Conversion and Management* 182 (February 2019): 72–88, https://doi.org/10.1016/j.enconman.2018.12.080.
27 Peter Singer, 'The Hinge of History,' Project Syndicate, October 8, 2021, www.project-syndicate.org/commentary/ethical-implications-of-focusing-on-extinction-risk-by-peter-singer-2021-10.
28 David C. Korten, *Change the Story, Change the Future* (Oakland CA: Berrett-Koehler Publishers, 2015), 34.
29 Korten, *Change the Story, Change the Future*, 29.
30 Korten, *Change the Story, Change the Future*, 25.
31 Klein, *This Changes Everything*, 21.
32 Korten, *Change the Story, Change the Future*, 132.

33 Korten, *Change the Story, Change the Future*, 74.
34 Lovelock, *The Revenge of Gaia*, 10.
35 Charles Percy Snow, *The Two Cultures* (London: Cambridge University Press, 1959).
36 Oreskcs and Conway, *The Collapse of Western Civilisation: A View from the Future*.
37 Lovelock, *The Revenge of Gaia*.
38 Sam Knights, 'Introduction: The Story so Far,' in *This Is Not a Drill: The Extinction Rebellion Handbook*, ed. Clare Farrell et al. (London: Penguin, 2019), 12.
39 Call me a luddite, but I have little faith in a future for unmanned ocean-going ships.
40 Sennett, *Together*.
41 Paul Raekstad and Sofa Gradin, *Prefigurative Politics: Building Tomorrow Today* (Cambridge; Medford, MA: Polity, 2020).
42 Ignaas Devisch, 'Een Realistisch Plan is Niet Realistisch,' *De Standaard*, November 9, 2021, www.standaard.be/cnt/dmf20211108_98016205.
43 Sennett, *Together*.

Bibliography

All URLs were accessible at the time of writing, unless otherwise stated.

Abbasov, Faïg. 'World's First "Carbon Neutral" Ship Will Rely on Dead-End Fuel.' Transport & Environment, February 23, 2021. www.transportenvironment.org/discover/worlds-first-carbon-neutral-ship-will-rely-dead-end-fuel/.

Adamopoulos, Anastassios. 'Maersk and CMA CGM Push EU for Free Emission Allowances.' *Lloyd's List*, November 30, 2020. https://lloydslist.maritimeintelligence.informa.com/LL1134910/Maersk-and-CMA-CGM-push-EU-for-free-emission-allowances.

——. 'Marshall Islands Demands Carbon Pricing Measure from the IMO.' *Lloyd's List*, November 16, 2020. https://lloydslist.maritimeintelligence.informa.com/LL1134705/Marshall-Islands-demands-carbon-pricing-measure-from-the-IMO.

——. 'Skou Pitches Minimum $450 per Tonne of Fuel Tax on Ships in the "Medium Term."' *Lloyd's List*, June 2, 2021. https://lloydslist.maritimeintelligence.informa.com/LL1136996/Skou-pitches-minimum-$450-per-tonne-of-fuel-tax-on-ships-in-the-medium-term.

Akenji, Lewis, Magnus Bengtsson, Viivi Toivio, Michael Lettenmeier, Tina Fawcett, Yael Parag, Yamina Saheb, et al. *1.5-Degree Lifestyles: Towards A Fair Consumption Space for All.* Berlin: Hot or Cool Institute, 2021. https://hotorcool.org/wp-content/uploads/2021/10/Hot_or_Cool_1_5_lifestyles_FULL_REPORT_AND_ANNEX_B.pdf.

Albion, Robert Greenhalgh. *Forests and Sea Power: The Timber Problem of the Royal Navy*. Cambridge, MA: Harvard University Press, 1926.

——. 'The Timber Problem of the Royal Navy 1652–1862.' *The Mariner's Mirror* 38, no. 1 (January 1952): 4–22. https://doi.org/10.1080/00253359.1952.10658102.

Bibliography

Anthropocene Working Group. 'Results of Binding Vote by AWG,' May 21, 2019. http://quaternary.stratigraphy.org/working-groups/anthropocene/.

Apuzzo, Matt, and Sarah Hurtes. 'Tasked to Fight Climate Change, a Secretive U.N. Agency Does the Opposite.' *New York Times*, June 3, 2021. www.nytimes.com/2021/06/03/world/europe/climate-change-un-international-maritime-organization.html.

Arendt, Hannah. *The Human Condition*. Second edition. Chicago IL: University of Chicago Press, 1998.

Arrhenius, Svante. 'On the Influence of Carbonic Acid in the Air upon the Temperature of the Ground.' *The London, Edinburgh, and Dublin Philosophical Magazine and Journal of Science* 41, no. 251 (April 1896): 237–76. https://doi.org/10.1080/14786449608620846.

Asafu-Adjaye, John, Linus Blomqvist, Stewart Brand, Barry Brook, Ruth DeFries, Erle Ellis, Christopher Foreman, et al. 'An Ecomodernist Manifesto.' Breakthrough Institute, 2015. http://rgdoi.net/10.13140/RG.2.1.1974.0646.

Australian Government. 'Australia's Long-Term Emissions Reduction Plan: A Whole-of-Economy Plan to Achieve Net Zero Emissions by 2050.' Canberra: Australian Government, 2020. www.industry.gov.au/sites/default/files/October%202021/document/australias-long-term-emissions-reduction-plan.pdf.

——. 'The Plan to Deliver Net Zero: The Australian Way.' Canberra: Australian Government, 2021. www.industry.gov.au/sites/default/files/October%202021/document/the-plan-to-deliver-net-zero-the-australian-way.pdf.

Balcombe, Paul, James Brierley, Chester Lewis, Line Skatvedt, Jamie Speirs, Adam Hawkes, and Iain Staffell. 'How to Decarbonise International Shipping: Options for Fuels, Technologies and Policies.' *Energy Conversion and Management* 182 (February 2019): 72–88. https://doi.org/10.1016/j.enconman.2018.12.080.

Bales, Kevin. 'Slavery in Its Contemporary Manifestations.' In *Critical Readings on Global Slavery*, edited by Damian Alan Pargas and Felicia Roşu, 1660–86. Leiden: Brill, 2018. https://doi.org/10.1163/9789004346611.

Bangstad, Sindre, and Torbjørn Tumyr Nilsen. 'Thoughts on the Planetary: An Interview with Achille Mbembe.' New Frame, September 5, 2019. www.newframe.com/thoughts-on-the-planetary-an-interview-with-achille-mbembe/.

Bannon, Eoin. 'UN Shipping Agency Greenlights a Decade of Rising Greenhouse Gas Emissions.' Transport & Environment, November 17, 2020. www.transportenvironment.org/press/un-shipping-agency-greenlights-decade-rising-greenhouse-gas-emissions.

Barns-Graham, William. 'Shipping Giant CMA CGN Reports Tenfold Profit Increase with Freight Rates Still Soaring.' The Institute of Export and International Trade, November 2021. www.export.org.uk/news/587994/Shipping-giant-CMA-CGN-reports-tenfold-profit-increase-with-freight-rates-still-soaring.htm.

Baucom, Ian. *Specters of the Atlantic: Finance Capital, Slavery, and the Philosophy of History*. Durham, NC: Duke University Press, 2005.

Baumler, Raphael. 'Working Time Limits at Sea, a Hundred-Year Construction.' *Marine Policy*, August 2020, 104101. https://doi.org/10.1016/j.marpol.2020.104101.

BBC. 'John Kerry Says Glasgow COP26 is the Last Best Hope for the World.' *BBC News*, October 19, 2021, sec. Scotland. www.bbc.com/news/uk-scotland-58914524.

Beck, Ulrich. *The Metamorphosis of the World*. Cambridge; Malden, MA: Polity, 2016.

——. *World at Risk*. Cambridge; Malden, MA: Polity Press, 2014.

Berrill, Paul. 'Sail Cargo Reveals Plans for Six 100-TEU Emission-Free Ships.' TradeWinds, November 1, 2021. www.tradewindsnews.com/technology/sailcargo-reveals-plans-for-six-100-teu-emission-free-ships/2–1–1091765.

Berry, Rhys. 'Wind-Propelled Liner to Guarantee Fuel Savings, Price Certainty, Says Neoline CEO.' *Bunkerspot*, June 7, 2021. www.bunkerspot.com/global/53387-global-wind-propelled-liner-to-guarantee-fuel-savings-price-certainty-says-neoline-ceo#.YL3yH7Ql614.linkedin.

Beyer, Jonny, Hilde C. Trannum, Torgeir Bakke, Peter V. Hodson, and Tracy K. Collier. 'Environmental Effects of the Deepwater Horizon Oil Spill: A Review.' *Marine Pollution Bulletin* 110, no. 1 (September 2016): 28–51. https://doi.org/10.1016/j.marpolbul.2016.06.027.

Binskin, Mark, Annabelle Bennett, and Andrew Macintosh. 'Royal Commission into National Natural Disaster Arrangements.' Canberra: Royal Commission into National Natural Disaster Arrangements, 2020.

Bjerg, Ole. *Parallax of Growth: The Philosophy of Ecology and Economy*. Cambridge; Malden, MA: Polity, 2016.

Blenkey, Nick. 'Eastern Pacific: "We Won't Carry Coal."' *MarineLog*, January 25, 2022. www.marinelog.com/legal-safety/environment/eastern-pacific-we-wont-carry-coal/.

Boers, Niklas. 'Observation-Based Early-Warning Signals for a Collapse of the Atlantic Meridional Overturning Circulation.' *Nature Climate Change* 11, no. 8 (August 2021): 680–8. https://doi.org/10.1038/s41558-021-01097-4.

Brum, Tony de. 'Minister de Brum of Marshall Islands Calls for Massive Global Response to Climate Crisis.' *Climate & Development Knowledge Network*, 2015. https://cdkn.org/2015/10/news-minister-de-brum-of-marshall-islands-calls-for-global-crisis-response-to-climate-change/?loclang=en_gb.

Bullock, Simon, James Mason, and Alice Larkin. 'The Urgent Case for Stronger Climate Targets for International Shipping.' *Climate Policy* 22, no. 3 (2022): 301–9. https://doi.org/10.1080/14693062.2021.1991876.

Bullock, Simon, James Mason, John Broderick, and Alice Larkin. 'Shipping and the Paris Climate Agreement: A Focus on Committed Emissions.' *BMC Energy* 2, no. 1 (December 2020): 5. https://doi.org/10.1186/s42500-020-00015-2.

Burkhart, Corinna, Matthias Schmelzer, and Nina Treu, eds. *Degrowth in Movement(s): Exploring Pathways for Transformation.* Winchester; Washington, DC: Zero Books, 2020.

Bush, Declan. 'EU States Block Zero-Emissions Resolution at IMO.' *Lloyd's List*, November 22, 2021. https://lloydslist.maritimeintelligence.informa.com/LL1138950/EU-states-block-zero-emissions-resolution-at-IMO.

——. 'Japan Backs Call to Make Shipping Net Zero by 2050.' *Lloyd's List*, October 26, 2021. https://lloydslist.maritimeintelligence.informa.com/LL1138623/Japan-backs-call-to-make-shipping-net-zero-by-2050.

——. 'Little Time to Talk Carbon Taxes as IMO States Wrestle with 2050 Resolution.' *Lloyd's List*, November 23, 2021. https://lloydslist.maritimeintelligence.informa.com/LL1138970/Little-time-to-talk-carbon-taxes-as-IMO-states-wrestle-with-2050-resolution.

Cambers, Sam. 'More than a Quarter of All Tonnage under Construction Will Use Alternative Fuels.' Splash247, November 30, 2020. https://splash247.com/more-than-a-quarter-of-all-tonnage-under-construction-will-use-alternative-fuels/.

Campling, Liam, and Alejandro Colás. *Capitalism and the Sea: The Maritime Factor in the Making of the Modern World.* London: Verso, 2021.

Carrington, Damian. '"Blah, Blah, Blah": Greta Thunberg Lambasts Leaders over Climate Crisis.' *The Guardian*, September 28, 2021, sec. Environment. www.theguardian.com/environment/2021/sep/28/blah-greta-thunberg-leaders-climate-crisis-co2-emissions.

Casarino, Cesare. *Modernity at Sea: Melville, Marx, Conrad in Crisis.* Theory out of Bounds 21. Minneapolis, MN: University of Minnesota Press, 2002.

Cheng, Lijing, John Abraham, Kevin E. Trenberth, John Fasullo, Tim Boyer, Ricardo Locarnini, Bin Zhang, et al. 'Upper Ocean Temperatures Hit Record High in 2020.' *Advances in Atmospheric Sciences* 38, no. 4 (April 2021): 523–30. https://doi.org/10.1007/s00376-021-0447-x.

Chircop, Aldo, and Desai Shan. 'Governance of International Shipping in the Era of Decarbonisation: New Challenges for the IMO?' In *Maritime Law in Motion*, edited by Proshanto K. Mukherjee, Maximo Q. Mejia, and Jingjing Xu, vol. 8:97–113. WMU Studies in Maritime Affairs. Cham: Springer International Publishing, 2020. https://doi.org/10.1007/978-3-030-31749-2_6.

Christoff, Peter. 'Net Zero by 2050? Even if Scott Morrison Gets the Nationals on Board, Hold the Applause.' *The Conversation*, June 23, 2021. http://theconversation.com/net-zero-by-2050-even-if-scott-morrison-gets-the-nationals-on-board-hold-the-applause-163074.

Clarke, Melissa. 'Australia Agreed to a Global Pact Reducing Fossil Fuels but the PM is Still Spruiking the Future of the Coal Industry.' *ABC News*, November 15, 2021. www.abc.net.au/news/2021-11-15/cop26-phase-down-coal-agreement-not-end-of-industry-says-pm/100621242.

Climate Action Tracker. 'Climate Action Tracker,' 2022. https://climateactiontracker.org/.

——. 'Climate Summit Momentum: Paris Commitments Improved Warming Estimate to 2.4°C.' Berlin: Climate Action Tracker, 2021. https://climateactiontracker.org/documents/853/CAT_2021-05-04_Briefing_Global-Update_Climate-Summit-Momentum.pdf.

——. 'International Shipping,' 2022. https://climateactiontracker.org/sectors/shipping/.

CMA CGM. 'The CMA CGM Group Heads towards Carbon Neutrality by 2050,' June 3, 2020. www.cma-cgm.com/news/3143/the-cma-cgm-group-heads-towards-carbon-neutrality-by-2050.

Coalition for Zero Emission Vessels. 'Leading Cargo Owners Stand Together for Maritime Decarbonization: Our Shared Ambition and Call for Policy Action.' Aspen Institute, 2021. www.cozev.org/img/FINAL-coZEV-2040-Ambition-Statement_2021-10-18-144834_uorz.pdf.

Cole, Laura. 'COP26: Commission Official Says Maritime Emissions Belong in NDCs.' ENDS Europe, November 2021. www.endseurope.com/article/1733097?utm_source=website&utm_medium=social.

Corbett, Jack, Mélodie Ruwet, Yi-Chong Xu, and Patrick Weller. 'Climate Governance, Policy Entrepreneurs and Small States: Explaining Policy Change at the International Maritime Organisation.' *Environmental Politics* 29, no. 5 (July 28, 2020): 825–44. https://doi.org/10.1080/09644016.2019.1705057.

Corbett, James J., James J. Winebrake, Erin H. Green, Prasad Kasibhatla, Veronika Eyring, and Axel Lauer. 'Mortality from Ship Emissions: A Global Assessment.' *Environmental Science & Technology* 41, no. 24 (December 1, 2007): 8512–18. https://doi.org/10.1021/es071686z.

Cowen, Deborah. *The Deadly Life of Logistics: Mapping Violence in Global Trade.* Minneapolis, MN: University of Minnesota Press, 2014.

Crowley, Kate. 'Is Australia Faking It? The Kyoto Protocol and the Greenhouse Policy Challenge.' *Global Environmental Politics* 7, no. 4 (November 2007): 118–39. https://doi.org/10.1162/glep. 2007.7.4.118.

Curry, Andrew. 'Revisiting The Limits to Growth.' *The Next Wave* (blog), August 13, 2021. https://thenextwavefutures.wordpress.com/2021/08/13/revisiting-the-limits-to-growth-systems-climate-change/.

Curry, Patrick. *Ecological Ethics: An Introduction.* Cambridge: Polity Press, 2011.

Cutcher, Nicola. 'Winds of Trade: Passage to Zero-Emission Shipping.' *American Journal of Economics and Sociology* 79, no. 3 (May 2020): 967–79. https://doi.org/10.1111/ajes.12331.

D'Alisa, Giacomo, Federico Demaria, and Giorgos Kallis, eds. *Degrowth: A Vocabulary for a New Era.* New York: Routledge, 2015.

Dana, Richard Henry Jr. *Two Years Before the Mast.* New York: Harper & Brothers, 1840.

Danowski, Déborah, and Eduardo Batalha Viveiros de Castro. *The Ends of the World.* Cambridge: Polity, 2017.

Davison, Nicola. 'The Anthropocene Epoch: Have We Entered a New Phase of Planetary History?' *The Guardian*, May 30, 2019, sec. Environment. www.theguardian.com/environment/2019/may/30/anthropocene-epoch-have-we-entered-a-new-phase-of-planetary-history.

De Beukelaer, Christiaan. 'Bored of Your Box Room? Try Being Marooned on the Ocean!' *Times Higher Education*, May 4, 2020. www.timeshighereducation.com/opinion/bored-your-box-room-try-being-marooned-ocean.

——. 'COVID-19 Border Closures Cause Humanitarian Crew Change Crisis at Sea.' *Marine Policy* 132 (October 2021). https://doi.org/10.1016/j.marpol.2021.104661.

——. 'COVID-19 at Sea: "The World as You Know it No Longer Exists."' *Cultural Studies* 35, no. 2–3 (May 4, 2021): 572–84. https://doi.org/10.1080/09502386.2021.1898020.

——. 'Feeling Flight Shame? Try Quitting Air Travel and Catch a Sail Boat.' *The Conversation*, October 2, 2019. https://theconversation.com/feeling-flight-shame-try-quitting-air-travel-and-catch-a-sail-boat-123349.

——. 'The Hundreds of Thousands of Stranded Maritime Workers Are the Invisible Victims of the Pandemic.' *Jacobin*, October 2020.

——. 'Tack to the Future: Is Wind Propulsion an Ecomodernist or Degrowth Way to Decarbonise Maritime Cargo Transport?' *Climate Policy* 22, no. 3 (2022): 310–19. https://doi.org/10.1080/14693062.2021.1989362.

De Decker, Kris. 'How to Design a Sailing Ship for the 21st Century?' *Low←Tech Magazine*, May 11, 2021. https://solar.lowtechmagazine.com/2021/05/how-to-design-a-sailing-ship-for-the-21st-century.html.

Delorme, Geoffroy. *L'homme-chevreuil: sept ans de vie sauvage.* Paris: les Arènes, 2021.

Demkes, Emy. 'The More Patagonia Rejects Consumerism, the More the Brand Sells.' The Correspondent, April 28, 2020. https://thecorrespondent.com/424/the-more-patagonia-rejects-consumerism-the-more-the-brand-sells.

Dempsey, Harry. 'Unvaccinated Sailors Risk Deepening Global Supply Chain Crisis.' *Financial Times*, March 22, 2021. www.ft.com/content/72feaabd-5f94-40ad-9073-0178db969b23.

Dempsey, Harry, and Dave Lee. 'Amazon, Ikea and Unilever Commit to Zero-Emission Shipping by 2040.' *Financial Times*, October 19, 2021. https://amp.ft.com/content/850eee4b-2c2d-4186-99d7-fdbe8131dddo.

Department of Industry, Science, Energy and Resources. 'National Greenhouse Gas Inventory Quarterly Update: June 2021.' Canberra: Australian Government, 2021. www.industry.gov.au/data-and-publications/national-greenhouse-gas-inventory-quarterly-update-june-2021.

DeSombre, Elizabeth R. *Flagging Standards*. Cambridge, MA: MIT Press, 2006.

——. *Why Good People Do Bad Environmental Things*. Oxford: Oxford University Press, 2018.

Devisch, Ignaas. 'Een Realistisch Plan is Niet Realistisch.' *De Standaard*, November 9, 2021. www.standaard.be/cnt/dmf20211108_98016205.

Dyke, James, Robert Watson, and Wolfgang Knorr. 'Climate Scientists: Concept of Net Zero is a Dangerous Trap.' *The Conversation*, April 22, 2021. http://theconversation.com/climate-scientists-concept-of-net-zero-is-a-dangerous-trap-157368.

Earsom, Joseph, and Tom Delreux. 'A Nice Tailwind: The EU's Goal Achievement at the IMO Initial Strategy.' *Politics and Governance* 9, no. 3 (September 30, 2021): 401–11. https://doi.org/10.17645/pag.v9i3.4296.

The Economist. 'Accelerating Energy Innovation for the Blue Economy.' *The Economist*, 2020. www.woi.economist.com/wp-content/uploads/2020/11/AcceleratingEnergyInnovationfortheBlueEconomy.pdf.

Ellis, Erle C. *Anthropocene: A Very Short Introduction*. Very Short Introductions. Oxford; New York: Oxford University Press, 2018.

Ellsmoor, James. 'Cruise Ship Pollution is Causing Serious Health and Environmental Problems.' Forbes, April 26, 2019. www.forbes.com/sites/jamesellsmoor/2019/04/26/cruise-ship-pollution-is-causing-serious-health-and-environmental-problems/.

Ėtkind, Aleksandr. *Nature's Evil: A Cultural History of Natural Resources*. Translated by Sara Jolly. New Russian Thought. Cambridge: Polity Press, 2021.

Etter, Lauren, and Brendan Murray. 'Shipping Companies Had a $150 Billion Year. Economists Warn They're Also Stoking Inflation.' Bloomberg, January 18, 2022. www.bloomberg.com/news/features/2022-01-18/supply-chain-crisis-helped-shipping-companies-reap-150-billion-in-2021.

European Commission. 'Delivering the European Green Deal.' Text. European Commission, 2021. https://ec.europa.eu/info/strategy/priorities-2019-2024/european-green-deal/delivering-european-green-deal_en.

Bibliography

European Environment Agency. 'Growth without Economic Growth.' Sustainability Transitions. Copenhagen: European Environment Agency, 2021. www.eea.europa.eu/themes/sustainability-transitions/drivers-of-change/growth-without-economic-growth.

Evans, Alan, Jessica Murray, Patrick Greenfield, Tom Levitt, Bibi van der Zee, and Samantha Lock. 'Cop26 President Declares "Fragile Win" for Climate despite Watered-down Coal Pledges – as It Happened.' *The Guardian*, November 14, 2021, sec. Environment. www.theguardian.com/environment/live/2021/nov/13/cop26-live-third-draft-text-expected-as-climate-talks-go-into-overtime.

Fagan, Brian M. *Beyond the Blue Horizon*. London: Bloomsbury Publishing, 2014.

Farand, Chloé. 'Anger as UN Body Approves Deal That Allows Ship Emissions to Rise to 2030.' Climate Home News, November 17, 2020. www.climatechangenews.com/2020/11/17/anger-un-body-approves-deal-allows-ship-emissions-rise-2030/.

Findlay, Helen S., and Carol Turley. 'Ocean Acidification and Climate Change.' In *Climate Change*, edited by Trevor M. Letcher 251–79. Elsevier, 2021. https://doi.org/10.1016/B978-0-12-821575-3.00013-X.

Fletcher, Max E. 'The Suez Canal and World Shipping, 1869–1914.' *The Journal of Economic History* 18, no. 4 (December 1958): 556–73. https://doi.org/10.1017/S0022050700107740.

Foucault, Michel. 'Des Espaces Autres.' Translated by Jay Miskowiec. *Architecture / Mouvement / Continuité*, no. 5, (October 1984): 46–9.

Franzen, Jonathan. *What If We Stopped Pretending?* London: 4th Estate, 2021.

Freestone, David, and Duygu Çiçek. 'Legal Dimensions of Sea Level Rise: Pacific Perspectives.' Washington DC: World Bank, 2021. https://openknowledge.worldbank.org/handle/10986/35881.

Gaffney, Owen, and Johan Rockström. *Breaking Boundaries: The Science of Our Planet*. London: DK, 2021.

George, Rose. *Ninety Percent of Everything: Inside Shipping, the Invisible Industry That Puts Clothes on Your Back, Gas in Your Car, and Food on Your Plate*. New York: Picador, 2014.

Ghosh, Amitav. *The Great Derangement: Climate Change and the Unthinkable*. Chicago, IL: University of Chicago Press, 2017.

——. *The Nutmeg's Curse*. London: John Murray, 2021.

Gilliam, Lucy. 'World Leaders Demand Shipping Emissions Cuts under Countries' Paris Climate Commitments.' Transport & Environment, December 13, 2017. www.transportenvironment.org/discover/world-leaders-demand-shipping-emissions-cuts-under-countries-paris-climate-commitments/.

Global Monitoring Laboratory. 'Mauna Loa CO2 Monthly Mean Data.' National Oceanic and Atmospheric Administration, 2022. https://gml.noaa.gov/webdata/ccgg/trends/co2/co2_mm_mlo.txt.

Gore, Tim. 'Confronting Carbon Inequality: Putting Climate Justice at the Heart of the COVID-19 Recovery.' Nairobi: Oxfam, 2020. www.oxfam.org/en/research/confronting-carbon-inequality.

Graeber, David. *Bullshit Jobs.* New York: Simon & Schuster, 2018.

Griffin, Michelle. 'Months at Sea with No Internet, Sailing Ship Heads Back to a "Different World."' *The Age,* May 1, 2020. www.theage.com.au/world/central-america/months-at-sea-with-no-internet-sailing-ship-heads-back-to-a-different-world-20200430-p540si.html.

Guterres, António. 'Our Common Agenda.' New York: United Nations, 2021.

——. 'Secretary-General Warns of Climate Emergency, Calling Intergovernmental Panel's Report "a File of Shame" While Saying Leaders "Are Lying", Fuelling Flames.' United Nations, April 4, 2022. www.un.org/press/en/2022/sgsm21228.doc.htm.

Hage, Ghassan. *Is Racism an Environmental Threat?* Cambridge; Malden, MA: Polity, 2017.

Hale, Thomas. 'Our Common Agenda – Governing the Future?' *Global Policy,* September 10, 2021. www.globalpolicyjournal.com/blog/10/09/2021/our-common-agenda-governing-future.

Hamilton, Clive, and Lins Vellen. 'Land-Use Change in Australia and the Kyoto Protocol.' *Environmental Science & Policy* 2, no. 2 (May 1999): 145–52. https://doi.org/10.1016/S1462-9011(99)000076.

Harari, Yuval Noah. *Sapiens: A Brief History of Humankind.* New York: Random House, 2015.

Haraway, Donna Jeanne. 'Anthropocene, Capitalocene, Plantationocene, Chthulucene: Making Kin.' *Environmental Humanities* 6, no. 1 (2015): 159–65. https://doi.org/10.1215/22011919-3615934.

Hardin, Garrett. 'The Tragedy of the Commons.' *Science* 162, no. 3859 (1968): 12438.

Harvey, Fiona. 'Campaigners Criticise Global Deal on Carbon Emissions from Shipping.' *The Guardian,* October 23, 2020, sec. Environment. www.theguardian.com/environment/2020/oct/23/green-groups-condemn-proposals-to-cut-shipping-emissions.

——. 'No New Oil, Gas or Coal Development if World is to Reach Net Zero by 2050, Says World Energy Body.' *The Guardian,* May 18, 2021, sec. Environment. www.theguardian.com/environment/2021/may/18/no-new-investment-in-fossil-fuels-demands-top-energy-economist.

——. 'Tony de Brum Obituary.' *The Guardian,* October 10, 2017, sec. Environment. www.theguardian.com/environment/2017/oct/10/tony-de-brum-obituary.

——. 'UN Chief Urges Airlines and Shipping Firms to Do More to Cut Emissions.' *The Guardian*, October 14, 2021, sec. Environment. www.theguardian.com/environment/2021/oct/14/un-chief-urges-airlines-and-shipping-firms-to-do-more-to-cut-emissions.

Hassan, Robert. *Uncontained: Digital Disconnection and the Experience of Time*. Melbourne: Grattan Street Press, 2019.

Heine, Hilda, and Christiana Figueres. 'Polluters on the High Seas.' *New York Times*, April 6, 2018. www.nytimes.com/2018/04/06/opinion/greenhouse-gases-international-shipping.html.

Hellenic Shipping News. 'EcoClipper Purchases Its First Sail Cargo Vessel,' January 21, 2022. www.hellenicshippingnews.com/ecoclipper-purchases-its-first-sail-cargo-vessel/.

Heller, Charles, and Lorenzo Pezzani. *The Left-to-Die Boat*. Documentary. Forensic Arcitecture, 2012. https://forensic-architecture.org/investigation/the-left-to-die-boat.

Hepburn, Sharon J. 'In Patagonia (Clothing): A Complicated Greenness.' *Fashion Theory* 17, no. 5 (November 2013): 623–45. https://doi.org/10.2752/175174113X13718320331035.

The Herald. 'Scotland to China and Back Again … Cod's 10,000-Mile Trip to Your Table.' *The Herald*. www.heraldscotland.com/default_content/12765981.scotland-china-back-cods-10-000-mile-trip-table/.

Herrington, Gaya. 'Update to Limits to Growth: Comparing the World3 Model with Empirical Data.' *Journal of Industrial Ecology* 25, no. 3 (June 2021): 614–26. https://doi.org/10.1111/jiec.13084.

Herwadkar, Nihar. 'Pros and Cons of Internet Onboard Ships: A Sailor's Perspective.' *Marine Insight* (blog), September 5, 2019. www.marineinsight.com/life-at-sea/seafaring-internet-onboard-ships-sailors-perspective/.

Hickel, Jason. *Less is More: How Degrowth Will Save the World*. London: Windmill Books, 2021.

Hilmola, Olli-Pekka. *Sulphur Cap in Maritime Supply Chains*. Basingstoke: Palgrave Pivot, 2019.

Hook, Andrew, Victor Court, Benjamin K. Sovacool, and Steve Sorrell. 'A Systematic Review of the Energy and Climate Impacts of Teleworking.' *Environmental Research Letters* 15, no. 9 (August 21, 2020): 093003. https://doi.org/10.1088/1748-9326/ab8a84.

Hulme, Mike. 'One Earth, Many Futures, No Destination.' In *Negotiating Climate Change in Crisis*, edited by Steffen Böhm and Sian Sullivan, 3–11. Cambridge: Open Book Publishers, 2021.

ICS. 'International Chamber of Shipping Sets out Plans for Global Carbon Levy to Expedite Industry Decarbonisation.' International Chamber of Shipping, September 6, 2021. www.ics-shipping.org/press-release/international-chamber-of-shipping-sets-out-plans-for-global-carbon-levy/.

ILO. 'Maritime Labour Convention.' Geneva: International Labour Organization, 2006.

IMO. 'Allow Crew Changes to Resolve Humanitarian Crisis, Insists IMO Secretary-General.' International Maritime Organization, September 7, 2020. www.imo.org/en/MediaCentre/PressBriefings/Pages/26-Allow-crew-changes.aspx.

——. 'Fourth IMO Greenhouse Gas Study.' London: IMO, 2020.

——. 'IMO Echoes Shipping Industry Calls for Governments to Keep Shipping and Supply Chains Open and Grant Special Travel Exemptions to Seafarers in COVID-19 Pandemic.' International Maritime Organization, March 31, 2020. www.imo.org/en/MediaCentre/PressBriefings/Pages/09-seafarers-COVID19.aspx.

——. 'IMO Ship Identification Number Scheme.' IMO, December 6, 2017. A 30/Res.1117. IMODOCS.

——. 'IMO Welcomes UN Resolution on Keyworker Seafarers.' International Maritime Organization, December 1, 2020. www.imo.org/en/MediaCentre/PressBriefings/pages/44-seafarers-UNGA-resolution.aspx.

——. 'Initial IMO Strategy on Reduction of GHG Emissions from Ships.' London: IMO, 2018.

——. 'Internet on Ships a Key to Recruiting and Retaining Seafarers, IMO Symposium Told,' September 25, 2015. https://imopublicsite.azurewebsites.net/en/MediaCentre/PressBriefings/Pages/40-WMD-symposium.aspx.

——. 'Third IMO Greenhouse Gas Study.' London: IMO, 2015.

InfluenceMap. 'Decision Time for the IMO on Climate: The Polarized Struggle among States for Ambitious Climate Policy on Shipping.' London: InfluenceMap, 2018. https://influencemap.org/site/data/000/309/IMO_Shipping_Report_April_2018.pdf.

International Energy Agency. 'Global CO2 Emissions Rebounded to Their Highest Level in History in 2021.' IEA, March 8, 2022. www.iea.org/news/global-co2-emissions-rebounded-to-their-highest-level-in-history-in-2021.

International Transport Forum. 'ITF Transport Outlook 2021.' Paris: OECD International Transport Forum, 2021.

International Transport Workers' Federation. 'ITF Agrees to Crew Contract Extensions.' International Transport Workers' Federation, March 19, 2020. www.itfglobal.org/en/news/itf-agrees-crew-contract-extensions.

——. 'UN Global Compact and Shipping Industry Confirm Formation of "People-Centred" Task Force to Ensure Just Transition to Net-Zero.' International Transport Workers' Federation, November 10, 2021. www.itfglobal.org/en/news/un-global-compact-and-shipping-industry-confirm-formation-people-centred-task-force-ensure.

IPCC. 'Climate Change 2022: Impacts, Adaptation and Vulnerability.' Intergovernmental Panel on Climate Change, 2022.

——. 'Climate Change 2022: Mitigation of Climate Change (Full Report).' Intergovernmental Panel on Climate Change, 2022.

——. 'Climate Change 2022: Mitigation of Climate Change (Summary for Policymakers).' Intergovernmental Panel on Climate Change, 2022.

——. 'Climate Change Widespread, Rapid, and Intensifying – IPCC.' *IPCC Newsroom*, August 9, 2021. www.ipcc.ch/2021/08/09/ar6-wg120210809-pr/.

——. 'The Evidence is Clear: The Time for Action is Now. We Can Halve Emissions by 2030. | UNFCCC,' 2022. https://unfccc.int/news/the-evidence-is-clear-the-time-for-action-is-now-we-can-halve-emissions-by-2030.

Jackson, Tim. *Post Growth: Life After Capitalism*. Cambridge: Polity, 2021.

——. *Prosperity without Growth: Foundations for the Economy of Tomorrow*. Second edition. London: Routledge, 2017.

James, C. L. R. *Mariners, Renegades, & Castaways: The Story of Herman Melville and the World We Live in; the Complete Text*. Reencounters with Colonialism: New Perspectives on the Americas. Hanover, NH: Dartmouth College, 2001.

Jepson, Paul, and Cain Blythe. *Rewilding: The Radical New Science of Ecological Recovery*. Hot Science. London: Icon Books, 2020.

Jiang, Jason. 'What Changes Would You like to Make to the Maritime Labour Convention?' Splash247, September 23, 2020. https://splash247.com/what-changes-would-you-like-to-see-the-maritime-labour-convention/.

Johnson, Taylor. 'Towards a Zero-Carbon Future.' *Mærsk*, June 26, 2019. www.maersk.com/news/articles/2019/06/26/towards-a-zero-carbon-future.

Kadir, Nazima. *The Autonomous Life? Paradoxes of Hierarchy and Authority in the Squatters Movement in Amsterdam*. Contemporary Anarchist Studies. Manchester: Manchester University Press, 2016.

Kallis, Giorgos. *The Case for Degrowth*. The Case for Series. Cambridge: Polity Press, 2020.

Kaplan, Jed O., Kristen M. Krumhardt, and Niklaus Zimmermann. 'The Prehistoric and Preindustrial Deforestation of Europe.' *Quaternary Science Reviews* 28, no. 27–8 (December 2009): 3016–34. https://doi.org/10.1016/j.quascirev.2009.09.028.

Kenner, Dario. *Carbon Inequality: The Role of the Richest in Climate Change*. Abingdon: Routledge, 2019.

Khalili, Laleh. *Sinews of War and Trade: Shipping and Capitalism in the Arabian Peninsula*. London; New York: Verso, 2020.

Klein, Naomi. *On Fire: The Burning Case for a Green New Deal*. London: Allen Lane, 2019.

——. *This Changes Everything: Capitalism vs. the Climate*. London: Penguin, 2015.

Knights, Sam. 'Introduction: The Story so Far.' In *This Is Not a Drill: The Extinction Rebellion Handbook*, edited by Clare Farrell, Alison Green, Sam Knights, and William Skeaping. London: Penguin, 2019.

Korten, David C. *Change the Story, Change the Future*. Oakland CA: Berrett-Koehler Publishers, 2015.

Krook, Joshua. 'Whatever Happened to the 15-Hour Workweek?' *The Conversation*, October 9, 2017. http://theconversation.com/whatever-happened-to-the-15-hour-workweek-84781.

Lafargue, Paul. *The Right To Be Lazy*. Chicago IL: Charles Kerr and Co, 1883.

Lasker, Phil, Jenya Goloubeva, and Bill Birtles. 'Here's How Australia is Planning to Deal with China's Ban on Foreign Waste.' *ABC News*, December 9, 2017. www.abc.net.au/news/2017-12-10/china-ban-on-foreign-rubbish-leaves-recycling-industry-in-a-mess/9243184.

Latouche, Serge. *Farewell to Growth*. Cambridge; Malden, MA: Polity, 2009.

Latour, Bruno. *After Lockdown*. Translated by Julie Rose. Cambridge: Polity Press, 2021.

——. *Down to Earth: Politics in the New Climatic Regime*. Cambridge; Medford, MA: Polity, 2018.

——. 'Imaginer les gestes-barrières contre le retour à la production d'avant-crise.' *AOC media – Analyse Opinion Critique* (blog), March 29, 2020. https://aoc.media/opinion/2020/03/29/imaginer-les-gestes-barrieres-contre-le-retour-a-la-production-davant-crise/.

Le Figaro. 'La Bretagne lance une filière de transport maritime à propulsion par le vent.' *Le Figaro*, November 10, 2021. www.lefigaro.fr/flash-eco/la-bretagne-lance-une-filiere-de-transport-maritime-a-propulsion-par-le-vent-20211110.

Leivestad, Hege Høyer, and Elisabeth Schober. 'Politics of Scale: Colossal Containerships and the Crisis in Global Shipping.' *Anthropology Today* 37, no. 3 (June 2021): 3–7. https://doi.org/10.1111/1467-8322.12650.

Lemaire, Tom. *De Val van Prometheus: Over de Keerzijden van de Vooruitgang*. Amsterdam: Ambo, 2010.

Levinson, Marc. *The Box: How the Shipping Container Made the World Smaller and the World Economy Bigger*. Second edition. Princeton, NJ: Princeton University Press, 2016.

Lewis, Simon L., and Mark A. Maslin. *The Human Planet: How We Created the Anthropocene*. London: Pelican Books, 2018.

Lloyd's Register and UMAS. *Zero-Emission Vessels 2030. How Do We Get There?* London: Lloyd's Register & University Maritime Advisory Services, 2017.

Lovelock, James. 'Beware: Gaia May Destroy Humans before We Destroy the Earth.' *The Guardian*, November 2, 2021, sec. Opinion.

www.theguardian.com/commentisfree/2021/nov/02/beware-gaia-theory-climate-crisis-earth.

——. *Gaia: A New Look at Life on Earth*. Second edition. Oxford Landmark Science. Oxford: Oxford University Press, 2016.

——. *The Revenge of Gaia*. London: Allan Lane, 2006.

Lyon, Christopher, Erin E. Saupe, Christopher J. Smith, Daniel J. Hill, Andrew P. Beckerman, Lindsay C. Stringer, Robert Marchant, et al. 'Climate Change Research and Action Must Look beyond 2100.' *Global Change Biology*, September 24, 2021, 1–13. https://doi.org/10.1111/gcb.15871.

Macfarlane, Robert. *Underland: A Deep Time Journey*. First American edition. New York: W. W. Norton & Company, 2019.

Mack, John. *The Sea: A Cultural History*. London: Reaktion Books, 2011.

Mærsk. 'A. P. Møller – Mærsk Accelerates Net Zero Emission Targets to 2040 and Sets Milestone 2030 Targets,' 2022. www.maersk.com/news/articles/2022/01/12/apmm-accelerates-net-zero-emission-targets-to-2040-and-sets-milestone-2030-targets.

Mærsk Mc-Kinney Møller Center for Zero Carbon Shipping. 'We Show the World it is Possible: Industry Transition Strategy October 2021.' Copenhagen: Mærsk Mc-Kinney Møller Center for Zero Carbon Shipping, October 2021.

Mair, Simon, Angela Druckman, and Tim Jackson. 'A Tale of Two Utopias: Work in a Post-Growth World.' *Ecological Economics* 173 (July 2020): 106653. https://doi.org/10.1016/j.ecolecon.2020.106653.

Mallinger, Kevin, and Martin Mergili. 'The Global Iron Industry and the Anthropocene.' *The Anthropocene Review*, December 30, 2020, 1–19. https://doi.org/10.1177/2053019620982332.

Malm, Andreas. *Fossil Capital: The Rise of Steam-Power and the Roots of Global Warming*. London: Verso, 2016.

——. *How to Blow up a Pipeline: Learning to Fight in a World on Fire*. London: Verso Books, 2021.

——. *The Progress of This Storm: Nature and Society in a Warming World*. London: Verso, 2017.

Malm, Andreas, and Alf Hornborg. 'The Geology of Mankind? A Critique of the Anthropocene Narrative.' *The Anthropocene Review* 1, no. 1 (April 2014): 62–9. https://doi.org/10.1177/2053019613516291.

Marine Benchmark. 'Maritime CO2 Emissions.' November 2020. Research Brief. Marine Benchmark, 2020. www.marinebenchmark.com/wp-content/uploads/2020/11/Marine-Benchmark-CO2.pdf.

The Maritime Executive. 'Unique Sail Cargo Ships Departs on First Atlantic Crossing From France.' The Maritime Executive, November 20, 2020. www.maritime-executive.com/article/unique-sail-cargo-ships-departs-on-first-atlantic-crossing-from-france.

Marshall Islands and Solomon Islands. 'Proposal for IMO to Establish a Universal Mandatory Greenhouse Gas Levy,' March 10, 2021. MEPC 76/7/12. IMODOCS.

Matthews, H. Damon, Katarzyna B. Tokarska, Joeri Rogelj, Christopher J. Smith, Andrew H. MacDougall, Karsten Haustein, Nadine Mengis, Sebastian Sippel, Piers M. Forster, and Reto Knutti. 'An Integrated Approach to Quantifying Uncertainties in the Remaining Carbon Budget.' *Communications Earth & Environment* 2, no. 1 (December 2021): 7. https://doi.org/10.1038/s43247-020-00064-9.

Mawani, Renisa. 'The Ship, the Slave, the Legal Person.' In *Studies in Law, Politics, and Society*, edited by Austin Sarat, George Pavlich, and Richard Mailey, 19–42. Bingley: Emerald Publishing Limited, 2022. https://doi.org/10.1108/S1059-433720220000087B002.

McCulloch, Ken. 'Sail Training.' In *Routledge International Handbook of Outdoor Studies*, edited by Barbara Humberstone, Heather Prince, and Karla A. Henderson, 236–43. New York: Routledge, 2015. https://doi.org/10.4324/9781315768465-27.

Mcdonald, Joshua. 'Rising Sea Levels Threaten Marshall Islands' Status as a Nation, World Bank Report Warns.' *The Guardian*, October 16, 2021, sec. World news. www.theguardian.com/world/2021/oct/17/rising-sea-levels-threaten-marshall-islands-status-as-a-nation-world-bank-report-warns.

McKibben, Bill. *Eaarth: Making a Life on a Tough New Planet.* New York: Time Books, 2010.

——. 'The Happiest Number I've Heard in Ages.' Substack newsletter. *The Crucial Years* (blog), January 7, 2022. https://billmckibben.substack.com/p/the-happiest-number-ive-heard-in.

——. 'It's Not Science Fiction.' *The New York Review*, December 17, 2020. www.nybooks.com/articles/2020/12/17/kim-stanley-robinson-not-science-fiction/.

Meadows, Donella H., Dennis L. Meadows, Jørgen Randers, and William W. Behrens. *The Limits to Growth.* New York: Universe Books, 1972.

Meinshausen, Malte, Jared Lewis, Christophe McGlade, Johannes Gütschow, Zebedee Nicholls, Rebecca Burdon, Laura Cozzi, and Bernd Hackmann. 'Realization of Paris Agreement Pledges May Limit Warming Just below 2 °C.' *Nature* 604, no. 7905 (April 14, 2022): 304–9. https://doi.org/10.1038/s41586-022-04553-z.

Merchant, Carolyn. *The Death of Nature: Women, Ecology, and the Scientific Revolution.* San Francisco: HarperOne, 1990.

Mills, M. Anthony, and Mark P. Mills. 'The Invention of the War Machine.' *The New Atlantis* 42, no. Spring (2014): 3–23.

Milne, Richard. 'Maersk Warns No End in Sight to Supply Chain Crisis as Profits Soar.' *Financial Times*, November 2, 2021. www.ft.com/content/2ede4d14-d0b2-4d85-9daf-420d3f14f9f7.

MOL. 'MOL, Tohoku Electric Power Sign Deal for Transport Using Coal Carrier Equipped with Hard Sail Wind Power Propulsion System (Wind Challenger).' Mitsui O.S.K. Lines, December 10, 2020. www.mol.co.jp/en/pr/2020/img/20085.pdf.

Monbiot, George. *Feral: Rewilding the Land, the Sea, and Human Life*. London: Penguin Books, 2014.

Monios, Jason, and Adolf K. Y. Ng. 'Competing Institutional Logics and Institutional Erosion in Environmental Governance of Maritime Transport.' *Journal of Transport Geography* 94 (June 2021): 103114. https://doi.org/10.1016/j.jtrangeo.2021.103114.

Monios, Jason, and Gordon Wilmsmeier. 'Maritime Governance after COVID-19: How Responses to Market Developments and Environmental Challenges Lead towards Degrowth.' *Maritime Economics & Logistics*, March 14, 2022. https://doi.org/10.1057/s41278-022-00226-w.

Morton, Adam, and Bec Pridham. 'Australia Considering More than 100 Fossil Fuel Projects That Could Produce 5% of Global Industrial Emissions.' *The Guardian*, November 2, 2021, sec. Environment. www.theguardian.com/environment/2021/nov/03/australia-considering-more-than-100-fossil-fuel-projects-that-could-produce-5-of-global-industrial-emissions.

Morton, Timothy. *Hyperobjects: Philosophy and Ecology after the End of the World*. Minneapolis, MN: University of Minnesota Press, 2013.

Myers, Norman. 'Environmental Refugees: A Growing Phenomenon of the 21st Century.' *Philosophical Transactions of the Royal Society of London. Series B: Biological Sciences* 357, no. 1420 (April 29, 2002): 609–13. https://doi.org/10.1098/rstb.2001.0953.

Nemra, Carsten Ned. 'Global Shipping is a Big Emitter, the Industry Must Commit to Drastic Action before it is Too Late.' *The Guardian*, September 19, 2021, sec. World news. www.theguardian.com/world/2021/sep/20/global-shipping-is-a-big-emitter-the-industry-must-commit-to-drastic-action-before-it-is-too-late.

Newby, Eric. *The Last Grain Race*. London: Secker & Warburg, 1956.

Nolan, Justine, and Martijn Boersma. *Addressing Modern Slavery*. Sydney: University of New South Wales Press, 2019.

Ocean Conservancy. 'UN Shipping Agency Greenlights a Decade of Rising Greenhouse Gas Emissions.' *Ocean Conservancy*, November 17, 2020. https://oceanconservancy.org/news/un-shipping-agency-greenlights-decade-rising-greenhouse-gas-emissions/.

Odijie, Michael. 'Cocoa and Child Slavery in West Africa.' In *Oxford Research Encyclopedia of African History*. Oxford University Press, 2020. https://doi.org/10.1093/acrefore/9780190277734.013.816.

Omidi, Maryam. 'Maldives Sends Climate SOS with Undersea Cabinet.' *Reuters*, October 17, 2009. www.reuters.com/article/us-maldives-environment-idUSTRE59G0P120091017.

Oreskes, Naomi, and Erik M. Conway. *Merchants of Doubt: How a Handful of Scientists Obscured the Truth on Issues from Tobacco Smoke to Global Warming*. New York: Bloomsbury Press, 2011.

——. *The Collapse of Western Civilisation: A View from the Future*. New York: Columbia University Press, 2014.

Osterkamp, Peder, Tristan Smith, and Kasper Søgaard. 'Five Percent Zero Emission Fuels by 2030 Needed for Paris-Aligned Shipping Decarbonization.' Getting to Zero Coalition, 2021. www.globalmaritimeforum.org/content/2021/03/Getting-to-Zero-Coalition_Five-percent-zero-emission-fuels-by-2030.pdf.

Ostrom, Elinor. *Governing the Commons: The Evolution of Institutions for Collective Action*. Cambridge: Cambridge University Press, 1990.

Paine, Lincoln P. *The Sea and Civilization: A Maritime History of the World*. London: Atlantic Books, 2015.

Parrique, Timothée. 'The Political Economy of Degrowth.' PhD, Université Clermont Auvergne & Stockholm University, 2019. https://tel.archives-ouvertes.fr/tel-02499463.

——. 'Sufficiency Means Degrowth,' 2022. https://timotheeparrique.com/sufficiency-means-degrowth/.

Patagonia. 'Don't Buy This Jacket, Black Friday and the New York Times,' November 25, 2011. https://eu.patagonia.com/gb/en/stories/dont-buy-this-jacket-black-friday-and-the-new-york-times/story-18615.html.

Pearce, Fred. 'How 16 Ships Create as Much Pollution as All the Cars in the World.' *Daily Mail*, November 11, 2009. www.dailymail.co.uk/sciencetech/article-1229857/How-16-ships-create-pollution-cars-world.html.

Pettifor, Ann. *The Case for the Green New Deal*. London: Verso, 2019.

Potter, Will. *Green Is the New Red: An Insider's Account of a Social Movement under Siege*. San Francisco: City Lights Books, 2011.

Raban, Jonathan. *Coasting: A Private Journey*. Vintage Departures. New York: Vintage Books, 2003.

Raekstad, Paul, and Sofa Gradin. *Prefigurative Politics: Building Tomorrow Today*. Cambridge; Medford, MA: Polity, 2020.

Ratcliffe, Rebecca, and Michael Standaert. 'China Coronavirus: Mayor of Wuhan Admits Mistakes.' *The Guardian*, January 27, 2020, sec. World news. www.theguardian.com/science/2020/jan/27/china-coronavirus-who-to-hold-special-meeting-in-beijing-as-death-toll-jumps.

Raworth, Kate. *Doughnut Economics: Seven Ways to Think like a 21st-Century Economist*. London: Random House Business Books, 2017.

Readfearn, Graham. '"Nothing off Limits": Offshore Gas and Oil Exploration Area 5km from Twelve Apostles.' *The Guardian*, June 15, 2021, sec. Australia news. www.theguardian.com/australia-news/2021/jun/15/nothing-off-limits-offshore-gas-and-oil-exploration-area-5km-from-twelve-apostles.

Rehmatulla, Nishatabbas, Sophia Parker, Tristan Smith, and Victoria Stulgis. 'Wind Technologies: Opportunities and Barriers to a Low Carbon Shipping Industry.' *Marine Policy* 75 (January 2017): 217–26. https://doi.org/10.1016/j.marpol.2015.12.021.

Rich, Nathaniel. *Losing Earth: The Decade We Could Have Stopped Climate Change.* London: Picador, 2020.

Robinson, Mary. *Climate Justice: Hope, Resilience, and the Fight for a Sustainable Future.* London: Bloomsbury, 2018.

Rockström, Johan, Will Steffen, Kevin Noone, Åsa Persson, F. Stuart Chapin, Eric F. Lambin, Timothy M. Lenton, et al. 'A Safe Operating Space for Humanity.' *Nature* 461, no. 7263 (September 2009): 472–5. https://doi.org/10.1038/461472a.

Rodrigue, Jean-Paul. 'Fuel Consumption by Containership Size and Speed.' The Geography of Transport Systems, 2020. https://transportgeography.org/contents/chapter4/transportation-and-energy/fuel-consumption-containerships/.

Rogelj, Joeri, Drew Shindell, Kejun Jiang, Solomone Fifita, Piers Forster, Veronika Ginzburg, Collins Handa, et al. 'Mitigation Pathways Compatible with 1.5°C in the Context of Sustainable Development.' Geneva: IPCC, 2018.

Rosling, Hans. *Factfulness: Ten Reasons We're Wrong about the World – and Why Things Are Better than You Think.* New York: Flatiron Books, 2018.

Russell, Joe. *The Last Schoonerman: The Remarkable Life of Captain Lou Kenedy.* Rockledge FL: Nautical Publishing Company, 2006.

Rutherford, Dan, Xiaoli Mao, and Bryan Comer. 'Potential CO_2 Reductions under the Energy Efficiency Existing Ship Index.' *International Council on Clean Transportation Working Paper* 27 (2020).

Ryan, Robert G., Jeremy D. Silver, and Robyn Schofield. 'Air Quality and Health Impact of 2019–20 Black Summer Megafires and COVID-19 Lockdown in Melbourne and Sydney, Australia.' *Environmental Pollution* 274 (April 2021): 116498. https://doi.org/10.1016/j.envpol.2021.116498.

Salleh, Ariel. *Ecofeminism as Politics: Nature, Marx and the Postmodern.* Second edition. London: Zed books, 2017.

Schijf, Manu, Pete Allison, and Kris Von Wald. 'Sail Training: A Systematic Review.' *Journal of Outdoor Recreation, Education, and Leadership* 9, no. 2 (2017): 167–80. https://doi.org/10.18666/JOREL-2017-V9-I2-8230.

Schultz, Colin. 'Shackleton Probably Never Took Out an Ad Seeking Men for a Hazardous Journey.' *Smithsonian Magazine.* www.smithsonianmag.com/smart-news/shackleton-probably-never-took-out-an-ad-seeking-men-for-a-hazardous-journey-5552379/.

Scutti, Susan. 'The Air Quality on Cruise Ships is so Bad, it Could Harm Your Health, Undercover Report Says.' CNN, January 26, 2019. www.cnn.com/2019/01/24/health/cruise-ship-air-quality-report/index.html.

Sekimizu, Koji. 'Future-Ready Shipping Conference 2015, Singapore.' International Maritime Organization, September 28, 2015. www.imo.org/en/MediaCentre/SecretaryGeneral/Pages/FRS-keynote.aspx.

Sekula, Allan, and Noël Burch. *The Forgotten Space*, 2010. www.theforgottenspace.net/.

Selin, Henrik, Yiqi Zhang, Rebecca Dunn, Noelle E. Selin, and Alexis K. H. Lau. 'Mitigation of CO_2 Emissions from International Shipping through National Allocation.' *Environmental Research Letters* 16, no. 4 (April 1, 2021): 045009. https://doi.org/10.1088/1748-9326/abec02.

Sennett, Richard. *Together: The Rituals, Pleasures and Politics of Cooperation*. London: Penguin Books, 2013.

Shachar, Ayelet. *The Birthright Lottery: Citizenship and Global Inequality*. Cambridge, MA: Harvard University Press, 2009.

Sharmina, Maria, O. Y. Edelenbosch, C. Wilson, R. Freeman, D. E. H. J. Gernaat, P. Gilbert, Alice Larkin, et al. 'Decarbonising the Critical Sectors of Aviation, Shipping, Road Freight and Industry to Limit Warming to 1.5–2°C.' *Climate Policy* 21, no. 4 (April 21, 2021): 455–74. https://doi.org/10.1080/14693062.2020.1831430.

Sharmina, Maria, Christophe McGlade, Paul Gilbert, and Alice Larkin. 'Global Energy Scenarios and Their Implications for Future Shipped Trade.' *Marine Policy* 84 (October 2017): 12–21. https://doi.org/10.1016/j.marpol.2017.06.025.

Sharpsteen, Bill. *The Docks*. Berkeley, CA: University of California Press, 2011.

Ship it Zero. 'Big Retail Fossil-Free Shipping Commitment Historic, But Too Weak.' Ship it Zero, October 19, 2021. https://shipitzero.org/climate-advocates-big-retail-fossil-free-shipping-commitment-historic-but-too-weak2/.

Shippingwatch. 'Nordic Minister Warns the Age of LNG-Fueled Ships is Over,' March 18, 2022. https://shippingwatch.com/carriers/article13839348.ece.

Simons, Andrew. 'Cargo Sailing: A Life Cycle Assessment Case Study.' Grasswil & Den Helder: 3SP Switzerland & EcoClipper, 2020.

Sims, R., R. Schaeffer, F. Creutzig, X. Cruz-Núñez, M. D'Agosto, D. Dimitriu, M. J. Figueroa Meza, et al. 'Transport.' In *Climate Change 2014: Mitigation of Climate Change. Contribution of Working Group III to the Fifth Assessment Report of the IPCC*, edited by O. Edenhofer, R. Pichs-Madruga, Y. Sokona, E. Farahani, S. Kadner, K. Seyboth, A. Adler, et al. Cambridge: Cambridge University Press, 2014.

Singer, Peter. 'The Hinge of History.' Project Syndicate, October 8, 2021. www.project-syndicate.org/commentary/ethical-implications-of-focusing-on-extinction-risk-by-peter-singer-2021-10.

Smil, Vaclav. *Energy and Civilization: A History*. Cambridge, MA: The MIT Press, 2017.

——. *Growth: From Microorganisms to Megacities*. Cambridge, MA: The MIT Press, 2019.

Smillie, Susan. 'Long Journey Home: The Stranded Sailboats in a Race to Beat the Hurricanes.' *The Guardian*, May 12, 2020. www.theguardian.com/environment/2020/may/12/long-journey-home-the-stranded-sailboats-in-a-race-to-beat-the-hurricanes.

Snow, Charles Percy. *The Two Cultures*. London: Cambridge University Press, 1959.

Soest, Heleen L. van, Michel G. J. den Elzen, and Detlef P. van Vuuren. 'Net-Zero Emission Targets for Major Emitting Countries Consistent with the Paris Agreement.' *Nature Communications* 12, no. 1 (December 2021): 2140. https://doi.org/10.1038/s41467-021-22294-x.

Steffen, Will, Johan Rockström, Katherine Richardson, Timothy M. Lenton, Carl Folke, Diana Liverman, Colin P. Summerhayes, et al. 'Trajectories of the Earth System in the Anthropocene.' *Proceedings of the National Academy of Sciences* 115, no. 33 (August 14, 2018): 8252–9. https://doi.org/10.1073/pnas.1810141115.

Stein, Sally. '"Back to the Drawing Board": Maritime Themes and Discursive Crosscurrents in the Notebooks of Allan Sekula.' In *OKEANOS*, by Allan Sekula, 60–89. Edited by Daniela Zyman and Cory Scozzari. Berlin: Sternberg Press, 2017.

Steinberg, Philip E. *The Social Construction of the Ocean*. Cambridge: Cambridge University Press, 2001.

Stern, Nicholas. *The Economics of Climate Change: The Stern Review*. Cambridge: Cambridge University Press, 2006.

Stockholm Environment Institute, International Institute for Sustainable Development, ODI, Third Generation Environmentalism, and United Nations Environment Programme. 'The Production Gap: Governments' Planned Fossil Fuel Production Remains Dangerously out of Sync with Paris Agreement Limits.' Stockholm: Stockholm Environment Institute, 2021. http://productiongap.org/2021report.

Stopford, Martin. *Maritime Economics*, Fourth edition. Hoboken: Taylor & Francis, 2013.

Symons, Jonathan. *Ecomodernism: Technology, Politics and the Climate Crisis*. Cambridge: Polity Press, 2019.

Taylor, Sarah McFarland. *Ecopiety*. Religion and Social Transformation. New York: NYU Press, 2019.

Thelen, Raphael. 'Ships in Mediterranean May Be Ignoring Refugees in Danger.' *Der Spiegel*, November 21, 2018, sec. International. www.spiegel.

de/international/europe/ships-in-mediterranean-may-be-ignoring-refugees-in-danger-a-1239495.html.

Timperley, Jocelyn. 'The Broken $100-Billion Promise of Climate Finance – and How to Fix It.' *Nature* 598, no. 7881 (October 21, 2021): 400–2. https://doi.org/10.1038/d41586-021-02846-3.

Tondo, Lorenzo. '"It's a Day Off": Wiretaps Show Mediterranean Migrants Were Left to Die.' *The Guardian*, April 16, 2021, sec. World news. www.theguardian.com/world/2021/apr/16/wiretaps-migrant-boats-italy-libya-coastguard-mediterranean.

Transparency International. *Governance at the International Maritime Organization*. Berlin: Transparency International, 2018.

Transport & Environment. 'Don't Sink Paris: Legal Basis for Inclusion of Aviation and Shipping Emissions in Paris Targets.' Brussels: Transport & Aviation, 2021. www.transportenvironment.org/wp-content/uploads/2021/10/Briefing-paper-NDCs-legal-advice-Aviation-Shipping-Final-2021-2.pdf.

——. 'Methane Escaping from "Green" Gas-Powered Ships Fuelling Climate Crisis – Investigation.' Transport & Environment, April 12, 2022. www.transportenvironment.org/discover/methane-escaping-from-green-gas-powered-ships-fuelling-climate-crisis-investigation/.

——. 'Planes and Ships Can't Escape Paris Climate Commitments,' May 4, 2018. www.transportenvironment.org/discover/planes-and-ships-cant-escape-paris-climate-commitments/.

Tree, Isabella. *Wilding: The Return of Nature to a British Farm*. London: Picador, 2019.

Trentmann, Frank. *Empire of Things: How We Became a World of Consumers, from the Fifteenth Century to the Twenty-First*. New York; London; Toronto: Harper Perennial, 2017.

Trilling, Daniel. 'How Rescuing Drowning Migrants Became a Crime.' *The Guardian*, September 22, 2020, www.theguardian.com/news/2020/sep/22/how-rescuing-drowning-migrants-became-a-crime-iuventa-salvini-italy.

Tsing, Anna Lowenhaupt. *Friction: An Ethnography of Global Connection*. Princeton, NJ: Princeton University Press, 2005.

Turner, Graham M. 'A Comparison of The Limits to Growth with 30 Years of Reality.' *Global Environmental Change* 18, no. 3 (August 2008): 397–411. https://doi.org/10.1016/j.gloenvcha.2008.05.001.

Turner, Victor W. *The Ritual Process: Structure and Anti-Structure*. The Lewis Henry Morgan Lectures 1966. New York: Aldine de Gruyter, 1995.

Ujifusa, Steven. *Barons of the Sea and Their Race to Build the World's Fastest Clipper Ship*. New York: Simon & Schuster, 2019.

UMAS and Getting to Zero Coalition. 'Closing the Gap: An Overview of the Policy Options to Close the Competitiveness Gap and Enable an

Equitable Zero-Emission Fuel Transition in Shipping.' Copenhagen: Getting to Zero Coalition, 2021.
UNCTAD. *Review of Maritime Transport.* Geneva: UNCTAD, 2020.
——. *Review of Maritime Transport.* Geneva: UNCTAD, 2021.
UNFCCC. 'Glasgow Climate Pact.' Conference of the Parties Serving as the Meeting of the Parties to the Paris Agreement. Glasgow: UNFCCC, 2021. https://unfccc.int/documents/311127.
UNHCR. 'UNHCR – Refugee Statistics.' UNHCR, 2022. www.unhcr.org/refugee-statistics/.
United Nations. 'Paris Agreement.' Paris: United Nations, 2015.
——. 'Rio Declaration on Environment and Development.' Rio De Janeiro: United Nations, 1992.
United States Coast Guard. 'On Scene Coordinator Report on Deepwater Horizon Oil Spill.' Washington DC: United States Coast Guard, 2011. https://homeport.uscg.mil/Lists/Content/Attachments/119/DeepwaterHorizonReport%20-31Aug2011%20-CD_2.pdf.
Urry, John. *Societies beyond Oil: Oil Dregs and Social Futures.* London; New York: Zed Books, 2013.
Vernick, Daniel. '3 Billion Animals Harmed by Australia's Fires.' World Wildlife Fund, July 28, 2020. www.worldwildlife.org/stories/3-billion-animals-harmed-by-australia-s-fires.
Viana, Mar, Pieter Hammingh, Augustin Colette, Xavier Querol, Bart Degraeuwe, Ina de Vlieger, and John van Aardenne. 'Impact of Maritime Transport Emissions on Coastal Air Quality in Europe.' *Atmospheric Environment* 90 (June 2014): 96–105. https://doi.org/10.1016/j.atmosenv.2014.03.046.
Vidas, Davor. 'The Law of the Sea for a New Epoch?' In *Tidalectics: Imagining an Oceanic Worldview through Art and Science*, edited by Stefanie Hessler, 231–40. London; Cambridge, MA: TBA21-Academy; The MIT Press, 2018.
Villiers, Alan. *The Last of the Wind Ships.* New York: Norton, 2000.
——. *Sons of Sindbad.* London: Hodder & Stoughton, 1940.
Vincent, Sam. *Blood & Guts: Dispatches from the Whale Wars.* Melbourne: Black Inc, 2014.
Wallace-Wells, David. *The Uninhabitable Earth: A Story of the Future.* London: Penguin Books, 2019.
Wang, Xiao-Tong, Huan Liu, Zhao-Feng Lv, Fan-Yuan Deng, Hai-Lian Xu, Li-Juan Qi, Meng-Shuang Shi, et al. 'Trade-Linked Shipping CO_2 Emissions.' *Nature Climate Change* 11, no. 11 (November 2021): 945–51. https://doi.org/10.1038/s41558-021-01176-6.
Weisman, Alan. *The World Without Us.* London: Virgin Books, 2004.
Weiss, Thomas G. *Would the World Be Better without the UN?* Cambridge: Polity Press, 2018.

Wiedmann, Thomas, Manfred Lenzen, Lorenz T. Keyßer, and Julia K. Steinberger. 'Scientists' Warning on Affluence.' *Nature Communications* 11, no. 1 (December 2020): 3107. https://doi.org/10.1038/s41467-020-16941-y.

Wilson, Edward O. *Half Earth*. New York: Liveright, 2016.

Wind-Assisted Ship Propulsion. 'Enkhuizen Nautical College Sets Course for Wind-Assisted Ship Propulsion and Sustainable Shipping.' *Wind-Assisted Ship Propulsion*, December 1, 2020. https://northsearegion.eu/wasp/news/enkhuizen-nautical-college-sets-course-for-wind-assisted-ship-propulsion-and-sustainable-shipping/.

——. 'New Wind Propulsion Technology: A Literature Review of Recent Adoptions.' Wind Assisted Ship Propulsion, 2020.

The World Bank. 'Carbon Revenues from International Shipping: Enabling an Effective and Equitable Energy Transition.' Washington DC: International Bank for Reconstruction and Development / The World Bank, 2022.

World Commission on Environment and Development. *Our Common Future*. Oxford Paperbacks. Oxford; New York: Oxford University Press, 1987.

Yamin, Farhana. 'The High Ambition Coalition.' In *Negotiating the Paris Agreement: The Insider Stories*, edited by Henrik Jepsen, Magnus Lundgren, Kai Monheim, and Hayley Walker, 216–44. Cambridge; New York: Cambridge University Press, 2021.

Zeldin, Theodore. *Conversation*. Mahwah, NJ: HiddenSpring, 2000.

Acknowledgements

When I embarked on the research project that led to this book, several years ago, I did not expect – forgive me the pun – smooth sailing. Given my academic background in musicology, cultural studies, and development studies, I am no stranger to shifting between epistemic registers and research topics. Even so, in my doctorate I explored the role of music industries in the socio-economic development of Burkina Faso and Ghana, which somehow combined my previous academic endeavours. Since completing that project, I have primarily looked at the broad topic of the interrelation between art and society, which became my academic bread and butter as a lecturer in cultural policy.

Nothing quite prepared me for what my change of tack, in terms of research focus, would bring. Nothing, apart from perhaps my keen interest in understanding how policies come to exist and what they do to the world. At the heart of my research lies, simply put, an interest in the tensions between 'the word and the world' – to invoke Arjun Appadurai's phrase in his book *Modernity at Large* (1996). Indeed, my research to date has always, in some way or another, tried to understand how policy serves as a field in which different interests and priorities come to a head. As such, my interest is not in policy as an abstract concept or a mere text resulting from a set of 'rational choices' as part of a transparent 'policy cycle.' Rather, I focus on policy as competing ideas, interventions, and contestations that aim to settle – or at least address – major issues. My work, which I believe to be best classified as 'anthropology of policy' regards policy-making as an ever-unfinished project influenced by people, power, and the inevitable (though never quite deterministic) path-dependency that emanates from history and context. In that sense, my research hasn't changed all that much, beyond an obvious shift in the object of my enquiry.

My 'work-life balance' is a long-lost cause. It tends towards an equilibrium as stable and realistic as 'net-zero' targets. It relies a lot on my not changing anything, while hoping that some *Deus ex Machina* will somehow balance things out. It is precisely the woefully inadequate imbalance between promised cuts in carbon emissions (which are, much like my 'life goals,' increasingly ambitious) and the ability to meet them (which are, much like my reduced focus on work, hardly noticeable in the aggregate).

I take great pleasure in the work I do and gladly accept the implications this commitment has for my personal life. The effort should, however, be justified by the challenge. That challenge turned out to be the decarbonisation of the shipping industry. Until 2017, it looked like the industry would continue to wholly reject any calls for climate action, which gave me a Cervantine impetus to tilt at wind ships. By exploring, in the first instance, the sense and non-sense of reimagining the potential of wind-propelled cargo ships, I decided to study the handful of initiatives that had emerged in the decade prior. I would apply my skills as an 'anthropologist of policy' – if I'm allowed to call myself that – to one of the least visible issues in climate action: maritime transport.

Meanwhile, sailing was meant to be the one regular thing that would get me away from work to think about – and do – something else. In 2016, shortly after relocating to Melbourne, I took up sailing in an attempt to definitively allocate every Saturday to something other than work. It was a success: I soon realised how much time on the water allowed me to reset and disconnect from the never-ending demands of academic labour. Switching off my phone and focusing on the horizon – and the craft of collaboration – allowed me to reclaim some of my life from work. A big thank you to Les and the crew of the *Windspeed* for teaching me jib from gybe and port from port. A special thanks to Greg H. and Geoff. It was in large part thanks to you – and everyone I've had the pleasure of sailing with on the *Tenacious*, the *Maybe*, the *Moosk*, and the *Excelsior* – who helped me (knowingly or not) to understand the craft and the history of sailing.

Needless to say, I quickly found a way to also bring sailing into the fold, though it took me a while to convince myself that I would be able to pursue this project in my existing role at the University of Melbourne – where I had thus far worked as a lecturer in cultural policy within an MA in Arts and Cultural Management. Many thanks as well to my programme colleagues Guy Morrow, Kirsten Stevens, Caitlin Vincent, Brian Long, Beth Driscoll, Ann Tonks, Kim Goodwin, and Kate MacNeill, for your understanding and support throughout the last few years. Many thanks to Miya Tokumitsu, Rachel Marsden, Gerhard Wiesenfeldt, Éliette Dupré-Husser, Nathan Bond, Grace Barrand, Kat Knights, Russell Goulbourne, Chris Healy, Charles Green, Frederik Vervaet, Meredith Martin, and many others within the university for offering listening ears and helpful comments.

Eventually, I started working on this new research project at a time that, it would later transpire, would be very difficult on a personal level. Weeks after my first brief article on sailing cargo vessels appeared in *The Conversation* ('Plain Sailing,' May 2018), which was within a month after the International Maritime Organization (IMO) agreed on its 'initial strategy' to tackle carbon emissions, I suffered a head injury, which left me with a lingering concussion. The slow recovery from this accident forced me to take a long period of sick leave during which I did not read, write, or think all that much. Keeping up to date with a research field that was rapidly evolving was not really an option; heading to sea for research the last of my worries. Luckily, I did recover at long last. When fully resuming work in January 2020, the prospect of fieldwork aboard a ship raised more than a few eyebrows. But if I was fully fit for work, so should I be able to head to sea, no? A good few risk assessments later, my Head of School and School Manager agreed that I should be able to continue my research project.

So, first of all, a big thank you to Charlotte Morgans and Jennifer Milam (and also Peter Otto) for being so supportive during my time away from work. Thank you for your understanding for my absence and your unrelenting (but gentle) encouragement to get back in the saddle.

A big thank you to Cornelius Bockermann of Timbercoast, the owner of the *Avontuur*, for welcoming me into your project. It's been a pleasure getting to know you and your ship, even if I did not expect to become so very closely acquainted with her. I hope that my story does justice to both the legacy of the *Avontuur* and your vision for her future.

Thank you to Captain Michael for being so open to my research project, when I joined the ship in Tenerife. Without your openness and support, my work would have been impossible. The biggest thanks of all are of course for the entire crew (Captain included) with whom I've had the pleasure to sail. The First Officer, Pol the Second Officer, Bo'sun, Giulia the Cook, Joni, Little Captain, Peggy, Mia, Pinkie, and of course Jennifer and Benji. A special thanks to Athena Corcoran-Tadd, who kindly agreed to relive a good part of our time together when drawing up the illustrations for this book. It's been a pleasure getting to know you all. Out at sea really was the best place to be.

Over the past years, I have been able to speak to a great variety of people involved in shipping, sailing, climate action, and thinking about the future. You are too many to list here, but I greatly appreciate the time you've all taken to speak to me.

Once I started writing, I benefited greatly from the feedback and encouragement from the numerous editors who offered comments on articles I've published in *Jacobin*, *The Monthly*, *Ocean*

Archive, *The Interpreter*, *De Standaard*, *The Conversation*, *Pursuit*, and others. Similarly, I would like to thank the (guest) editors of *Cultural Studies* (Ted Striphas and John Erni), *Marine Policy* (Quentin Hanich), and *Climate Policy* (notably Jan Corfee-Morlot) – as well as the anonymous reviewers enlisted by them – for their helpful comments on my articles in these journals.

This book wouldn't have been here without Tom Dark, my editor at Manchester University Press. Thank you for your enthusiasm about the project and your enduring support throughout the process of writing and editing. Beyond Tom's editorial suggestions, the manuscript improved markedly from the feedback so kindly provided by Brian Long, who kindly read the first drafts of all chapters, Dr Lucy Gilliam, Jason Anderson, and Phil Steinberg – your insights in the shipping industry and the intricacies of the oceans and seas have helped me eliminate many an error from the book. Once the manuscript was fully written, I benefited from great help and support from the entire team at Manchester University Press, including Rebecca Parkinson, Humairaa Dudhwala, David Appleyard, Chris Hart, Christian Lea, and Jen Mellor. Also Rachel Evans (copy editor), Sarah Cook (index), and Louise McGuinness (proofreading) helped a great deal to improve the text and its appearance. Though, rest assured, any remaining errors or omissions remain entirely my own.

During the Finnish spring of 2022, I was able to revise the manuscript as a visitor of the University of Jyväskylä. Thanks to the unfragmented time and mental space I had there, in between fieldwork travels, I was able to refine and improve the book significantly. Many thanks to Miikka Pyykkönen for hosting me, and Olli-Pekka and Helmi of the *Keski-Suomen Kirjailijat* (the Writer's House of Central Finland), for kindly housing me during this period.

In case you were wondering, Jean François De Beukelaer, to whom the book is dedicated, was my late paternal grandfather – a sailor at heart, if not so much in practice. Family mythology has it that, when faced with the choice between the prospect of marrying my grandmother and the call of the sea, he chose the former. *Merci Papoe.*

Most of all, Genevieve, thank you for keeping me company throughout the voyage and being there upon my return to Australian shores.

Index

Note: 'f' after a page number indicates a figure on that page; 'n' indicates a note; 'p' indicates a photograph

Index

Index

Index

Index

Index

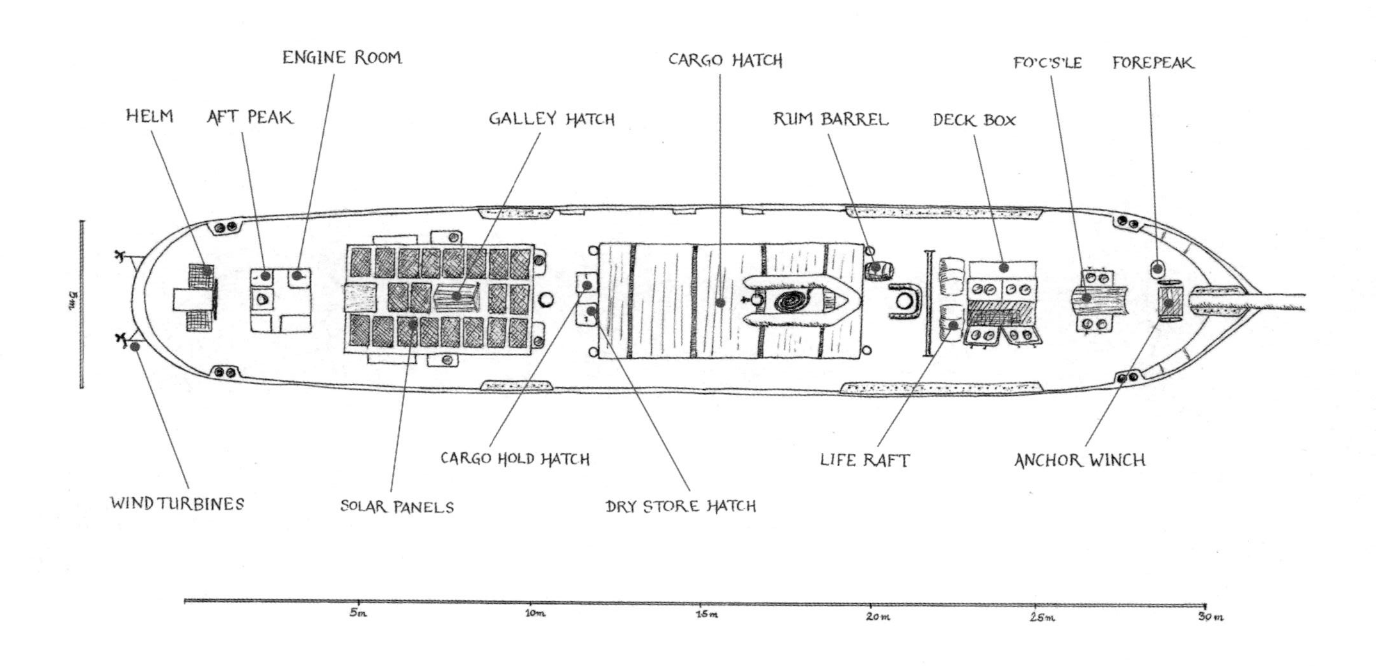

ENGINE ROOM
CARGO HATCH
FO'C'S'LE
FOREPEAK
HELM
AFT PEAK
GALLEY HATCH
RUM BARREL
DECK BOX
5m
CARGO HOLD HATCH
LIFE RAFT
ANCHOR WINCH
WIND TURBINES
SOLAR PANELS
DRY STORE HATCH
5m
10m
15m
20m
25m
30m

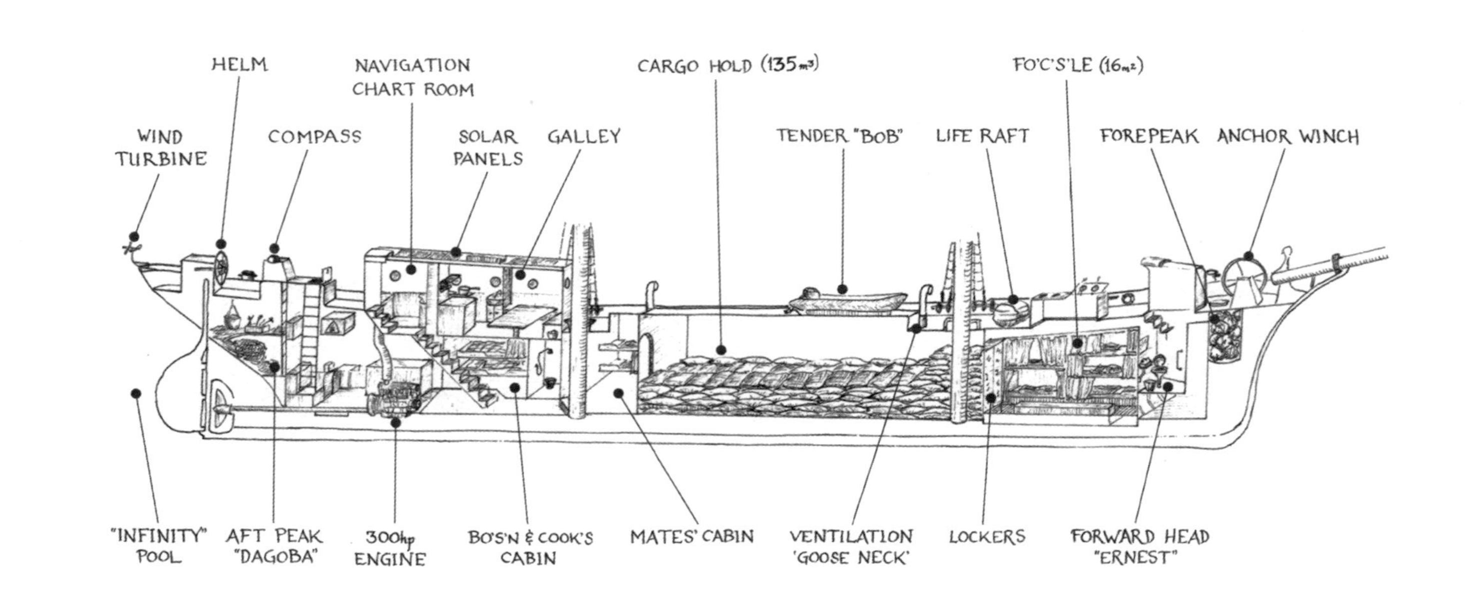
HELM
NAVIGATION CHART ROOM
CARGO HOLD (135m³)
FO'C'S'LE (16m²)
WIND TURBINE
COMPASS
SOLAR PANELS
GALLEY
TENDER "BOB"
LIFE RAFT
FOREPEAK
ANCHOR WINCH
"INFINITY" POOL
AFT PEAK "DAGOBA"
300hp ENGINE
BO'S'N & COOK'S CABIN
MATES' CABIN
VENTILATION 'GOOSE NECK'
LOCKERS
FORWARD HEAD "ERNEST"

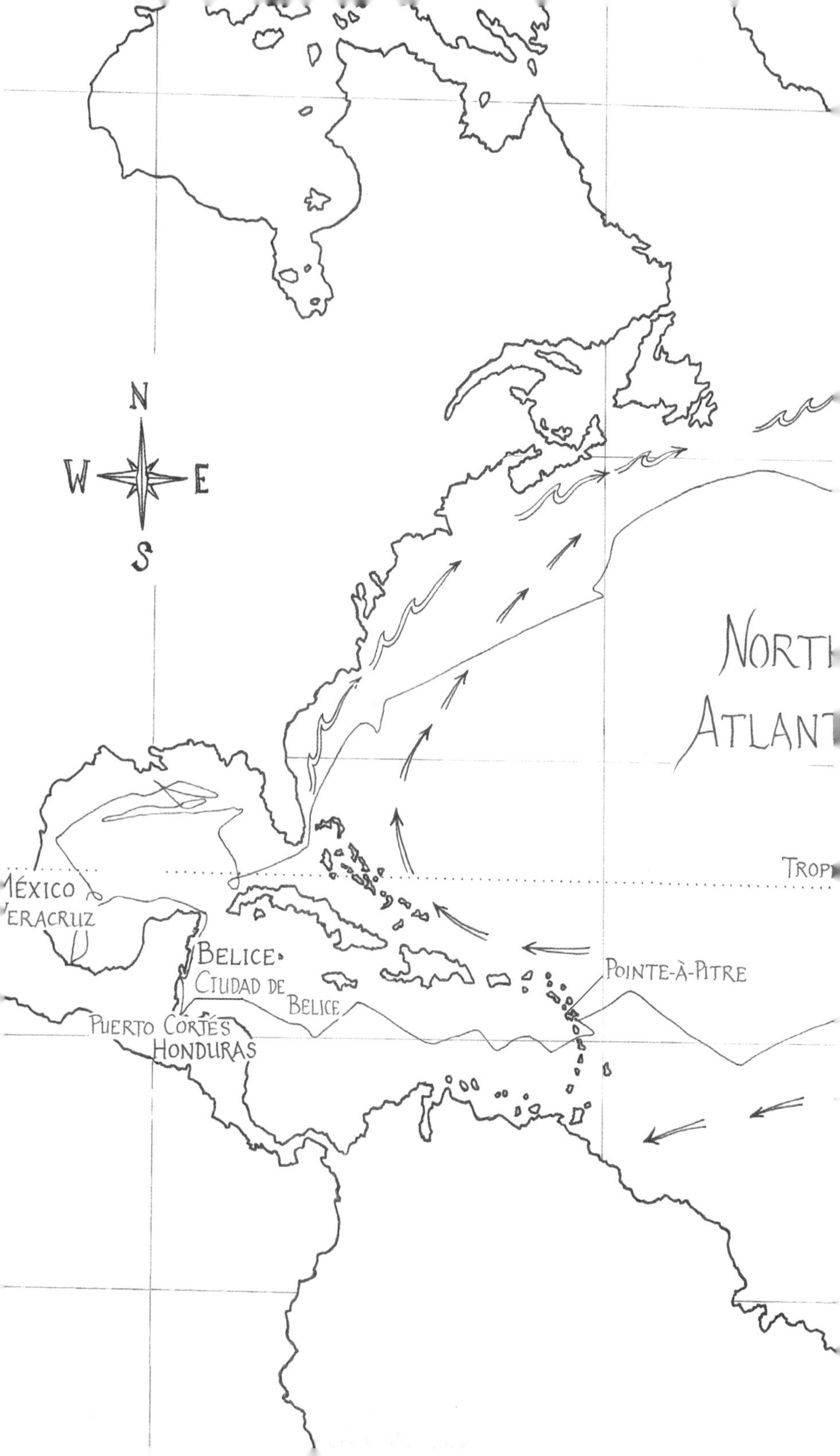

N
W
E
S
NORTH
ATLANT
TROP
MÉXICO
VERACRUZ
BELICE
CIUDAD DE
BELICE
POINTE-À-PITRE
PUERTO CORTÉS
HONDURAS